C000247153

MAPS AND HISTORY

IN SOUTH-WEST ENGLAND

EXETER STUDIES IN HISTORY
General Editor: Jonathan Barry

Other titles in this series include:

Other titles published by the University of Exeter Press

MAPS AND HISTORY IN SOUTH-WEST ENGLAND

Edited by

Katherine Barker and Roger J.P. Kain

on behalf of the University of Exeter Centre for South-Western Historical Studies

Exeter Studies in History No. 31
University of Exeter Press

First published in 1991 by
University of Exeter Press
Reed Hall
Streatham Drive
Exeter EX4 4QR
UK

© Katherine Barker, Roger J.P. Kain and the several authors each in respect of the paper contributed 1991

Typeset by Nigel Code, Department of History and Archaeology, University of Exeter

Printed in the UK by Short Run Press Ltd, Exeter

British Library Cataloguing in Publication Data
  Maps and history in South-West England
  I.Barker,Katherine II. Kain, Roger
  942.3

  ISBN 0-85989-373-1

Contents

List of Figures

The Contributors

Jennifer Baker (née Gambier) graduated from the University of Bristol and teaches Geography at Exeter College. She carried out her doctoral research in the Departments of Economic History and Geography at the University of Exeter on tithe commutation in Dorset.

Katherine Barker is a graduate of the University of Birmingham where she read archaeology and geography and took historical geography as a special subject with the late Professor Harry Thorpe. She is part-time tutor with Bristol University Department of Continuing Education and has published a number of papers on West Country history. She recently both contributed to, and edited *The Cerne Abbey Millenium Lectures* and is at present chairman of Dorset Museums Association and an Honorary Research Fellow in the University of Exeter Centre for South-Western Historical Studies.

John Chapman is a Principal Lecturer in Geography at Portsmouth Polytechnic. His research interests are in agricultural and land-use changes from the eighteenth century onwards, and more especially in parliamentary enclosure and its effects. He has published a number of articles on the subject, and his *Guide to the Parliamentary Enclosures of Wales* is due to appear shortly. He is currently working on non-parliamentary enclosures in southern England and on failed enclosure bills.

Graham Haslam received his doctorate in Tudor/Stuart history from Louisiana State University. From 1975 until 1991 he was employed by the Duchy of Cornwall as an archivist. He has recently joined The Yale Center for Parliamentary History, University of Yale, to work as an assistant editor.

Roger Kain is Professor of Geography, University of Exeter and a Fellow of the British Academy. He has written a number of books and articles on cadastral maps, especially tithe surveys, and is currently working with Richard Oliver on a descriptive and analytical catalogue of the tithe maps of England and Wales.

Richard Oliver is a graduate of the University of Sussex where he wrote his doctoral thesis on The Ordnance Survey in Great Britain 1835-70. He has published a number of articles on the Ordnance Survey and has collaborated with Professor Brian Harley on the Harry Margary Ordnance Survey Old Series facsimile project. Dr Oliver is editor of *Sheetlines,* the journal of the Charles Close Society for the Study of Ordnance Survey maps, and is Leverhulme Trust Research Fellow in the History of Cartography, University of Exeter.

William Ravenhill is Emeritus Reardon Smith Professor of Geography in the University of Exeter. He has written extensively on the archaeology and historical and regional geography of south-west England and, from the mid-1960s, Professor Ravenhill pioneered the scholarly study of the history of cartography in Devon and Cornwall. In 1965 the Devon and Cornwall Record Society and the University of Exeter published his *Benjamin Donn: A Map of the County of Devon, 1765* which was followed in 1972 by *John Norden's Manuscript Maps of Cornwall and its Nine Hundreds.* For the future, he is writing on English mapping for an international *History of Cartography,* and, nearer home, he is preparing a facsimile edition of Joel Gascoyne's 1699 *Map of Cornwall* for the Devon Record Society and is co-editor with Roger Kain of the Centre for South-Western Historical Studies' *Historical Atlas of South-West England.* Professor Ravenhill was the University's Harte Lecturer in Local History in 1990.

Preface

Maps and History

This collection of six essays on the theme of maps and history in South-West England is based on five papers read at a conference in Sherborne, Dorset which Katherine Barker convened in May 1990 on behalf of the Centre for South-Western Historical Studies. In addition it prints the text of Professor Ravenhill's 1990 University of Exeter Harte Lecture in Local History.

Maps are one of the oldest forms of human communication, not as old as speech but in many pre-historic societies graphicacy (the ability to communicate by maps) predates literacy. Although maps are relatively simple iconic devices, much of the research and writing about early maps has been concerned with the increasing accuracy with which maps could be made. Such studies focus on the ways in which developments in mathematics from the Renaissance onwards were applied to surveying, how scientific knowledge improved the instruments with which survey was accomplished, and on changes in the techniques by which survey data were converted into maps and then published. The essays in this book do not directly address such matters which more properly concern the history of the art and science of cartography. Our contributors are concerned more with the practical and political purposes for which maps were used, about the symbolic and ideological roles of maps in the history of South-West England, and about the ways in which map evidence can be used to retrieve facts about the past for use in the writing of history. Indeed a number of the different epistemologies of map history are represented in this book:

History of maps: an approach which views maps essentially as artefacts and is concerned with unravelling the origins of particular maps, dating them, finding out about those who constructed them and (with printed maps), of searching out and listing various editions in cartobibliographies, cataloguing and valuing them, and providing a service to collectors. Such are the concerns of Professor Ravenhill in the text of his 1990 Harte Lecture (Chapter 1).

Maps in history: assessing the roles and uses of maps by their contemporary and subsequent users, the knowledge that they stored and the power that knowledge imparted and the control over land and property that possession of maps provided. This focus on the social significance of the map in the past is very much the approach of Mrs Barker and Dr Haslam in their essays on estate mapping (Chapters 2 and 3).

History in maps: the idea that historical maps are repositories of information about the times at which they were compiled and so can be used to reconstruct components of the history and geography of those past times. It is a view which sees historical maps as historical sources and is espoused, for example, by Dr Chapman in his essay on parliamentary enclosure maps (Chapter 4) and by Professor Kain, Dr Oliver and Dr Baker with respect to the tithe surveys of the mid-nineteenth century (Chapter 5).

Historical cartography: the activity of constructing maps from data relating to the past. The Centre's *Historical Atlas of South-West England* is one such study.

History of cartography: narrowly defined this is the study of changes in the art, science and technology of map-making itself, the mathematics of projections, the science of land surveying, the use of conventional symbols on maps, the processes of reproduction etc. Dr Oliver takes us through the activities of the Ordnance Survey in South-West England in this manner (Chapter 6).

It is a basic contention of the contributors and editors of this book that maps are historical phenomena of great significance. They were important agents of change in history as well as being of interest as historical artefacts and of value as sources of evidence on the past. That is why the book is entitled: Maps *and* History.

Sherborne and Exeter
June 1991

Acknowledgement

We wish to thank Dr Jonathan Barry, General Editor of *Exeter Studies in History* for his most helpful and informed comments on our text.

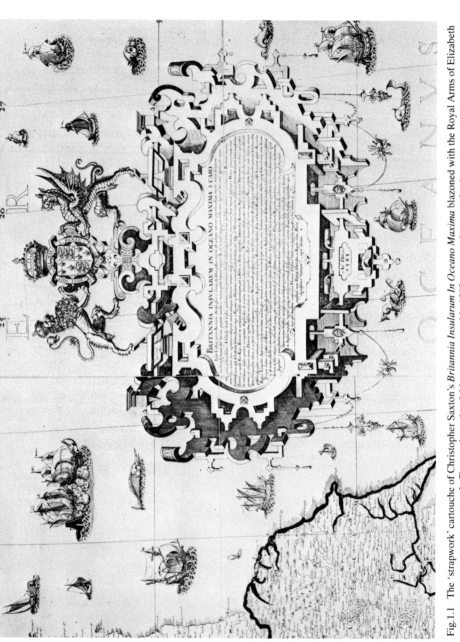

Fig.1.1 The 'strapwork' cartouche of Christopher Saxton's *Britannia Insularum In Oceano Maxima* blazoned with the Royal Arms of Elizabeth I. Two copies only of this map engraved in 1583 appear to have survived.

CHAPTER ONE

The South West in the Eighteenth-Century Re-mapping of England

(The Harte Lecture 1990)

William Ravenhill

The eighteenth century was marked by an unprecedented surge in county mapping on what was, for the time, the large scale of about one inch to the mile, undertaken by private cartographers. The century opens with the first of these, Joel Gascoyne, producing his map of Cornwall. This primacy for Cornwall and indeed for the South West is viewed in this essay against the backcloth of what preceded and what was to occur in the sphere of map-making in the rest of the country. This re-mapping of England falls quite neatly between the cartographies which emerged as the result of the output from two official structures separated by two hundred years. The later of these structures is, of course, the Ordnance Survey. As a mapping agency it stemmed from the combined involvement of the Crown, namely George III, the Royal Society, and the Board of Ordnance in the years after 1783.

In 1583, Christopher Saxton's large general map of England and Wales, blazoned with the royal arms, was printed (Figure 1.1), for its time an outstanding achievement, and again the output from an earlier 'official structure'.[1] Those two last words are between inverted commas, because in Elizabeth's reign we should not be looking for, or expecting, a clear-cut and tidy structure. Nevertheless, official it was, if appearing to us somewhat indirect, but this is the way Elizabeth's administration worked. Let me recall to your memories that Christopher Saxton was surveying under the direct sponsorship of Thomas Seckford, Master of Requests to Queen Elizabeth; also deeply involved, and considered the prime mover, was her Lord Treasurer, William Cecil, Lord Burghley. The 'placart' or pass allowing Saxton to travel and to gain access to any part of the realm was signed by ten members of the Privy Council which declared him to be the Queen's appointee. There is no doubt

1

that this was an official survey promoted by the Crown on the advice of ministers, as an act of policy, and designed to produce maps for the purpose of national administration and defence, for a state which was increasingly forced by circumstances to become more interventionist in parish and county affairs.

Survey and mapping of this magnitude, nationwide in space and supported by government and the Crown was not to be repeated for over two hundred years (Figure 1.2). The prime reason for this change of policy at the very centre of our national administration resulted, and this may surprise you, from Saxton's overkill of success, which was in the course of time to the embarrassment of the Crown. The maps fulfilled their intended function of meeting the changing needs of central government but they were also used to enhance the royal image of the Virgin Queen, Gloriana, the temple of Eliza and its realm. To examine the portraits of Elizabeth is to witness the creation of a legend and a cult with the deliberate aim of establishing a symbol designed to hold together a people not only deeply divided by social hierarchy and religious beliefs, but also threatened by external enemies.[2] The portraits endeavour to transmute her from an aristocratic lady into a cosmic vision hovering over Her England. Gheeraerts paints the petite royal toes on a rendering of Saxton's 1583 map (Figure 1.3). We bear witness to a highly developed propaganda machine endeavouring to secure victory over reason.

In case our readers in Cornwall feel left out may I assure them that the painting has been reduced on both sides; before trimming the original would have extended to Land's End, and its impact would have been even more geographically striking than this truncated relic. Elizabeth is England, Queen and Country are one. This famous map, and more so Saxton's *Atlas* did much to assist in the enrichment of this image. The royal arms are there on every sheet of the *Atlas* to proclaim that these are the Queen's maps, this is the Queen's land, her Kingdom and, lest the point be missed, there is the well-known frontispiece; it bears no title, no reference to Seckford, to Saxton or to anybody, only the Queen enthroned, surmounted by her arms and an emblem of her rule, flanked by figures representing Cosmography and Geography, underscored by verses celebrating the accomplishments of her benign reign. Recently, the extensive and sumptuous wardrobe of Elizabeth, and how she indulged in an almost barbaric display of rich fabrics and jewellery has been well documented.[3] Nevertheless, she is pictured in the frontispiece wearing robes reserved exclusively for the state opening of parliament; in her long reign of 44 years she held only ten parliaments, so the significance of their being worn for this purpose should be duly noted as symbolically linking the land as portrayed by the map with the Queen's administration.

Ah! but no ministers, no privy councillors, well-meaning, loyal and sincere though they may be, can foresee all the consequences to their successors of the sentiments they have helped to crystallise and for which they have provided a focus and an

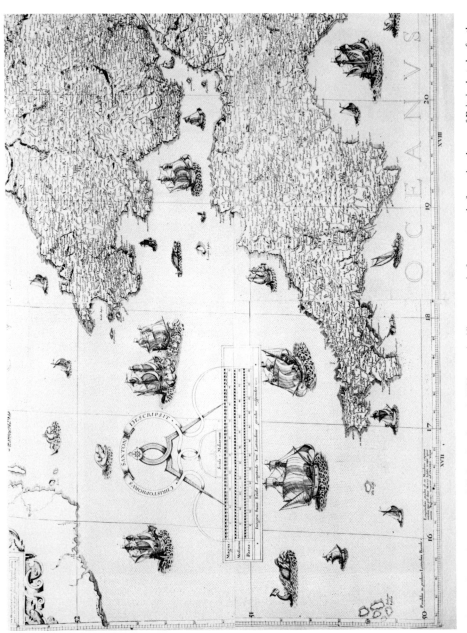

Fig.1.2 The south-west part of Christopher Saxton's 1583 map significant not only for a portrayal of a regional area of Britain, but also for the sophistication of the representation of scale and for its geographical location in relation to the global co-ordinate system. Note the statement in the extreme south-west corner relating the co-ordinates to a prime meridian running through the Azores.

Fig.1.3   The Ditchley portrait of Queen Elizabeth attributed to Marcus Gheeraerts the
Younger. The trimming of the picture on both sides is revealed by the truncation of the
sonnet on the right and the removal of a part of Cornwall in the west. The Queen is painted
as if standing on a rendering of Saxton's map of 1583. Photograph supplied by the National
Portrait Gallery London and reproduced by their kind permission.

emblem. Saxton's superb cartography, successful in its own right, while undoubtedly re-enforcing the royal cult at first, in due course with its wider dissemination began, however unintentionally, to compete with it. Cartography by its intrinsic graphic form emphasises the map content. It becomes the eye-catcher; symbols of royalty and patronage, make them as explicit as you may, must be relegated to the voids of the marginalia, merely decorative and in time dispensable. All men are of a Time, no doubt, but all men are also of a Place; the land began to tell. For the first time the English could take an effective visual and conceptual possession of the physical landscape in which they lived. For the first time, they could perceive in considerable detail, their country, their county, their town, their home, and pointing to a spot on the paper landscape, they could feel as well as declare 'Here, I belong'. The effect was as dramatic as it was deep.[4]

For a Queen, jealously guarding her court as the unique setting for majesty and as a focus for allegiance in the state, would hardly welcome being outshone by any other person or thing. Even so the challenge was there, its inexorable progress was difficult to impede. Saxton had already been given a privilege for ten years to print and sell his maps. What could be done to stem the tide of changed affections was done, namely withdraw any future royal support. When, therefore, John Norden came along with his scheme for a 'Re-description of England', his *Speculum Britanniae*, and a special volume for Cornwall in which he included individual maps for its nine hundreds, neither the royal coffers nor the regal smiles of encouragement of either Elizabeth or James were there to greet him.[5] And so it was to remain for a long time. If the work of mapping was to go on at all, it would have to be under other auspices and with other support. It did go on, but henceforth new forms of patronage and the market place dictated the pace and, importantly, the space; the main responsibility resting with the map-trade and with a narrowly based map-buying public who had to be wooed as subscribers.

Saxton's copper plates had a long life; they were still being printed from after two centuries of use.[6] Following issues of the *Atlas* when Saxton retained his 'privilege', that is monopoly, there were no less than eight definite editions, the last being in 1770. Plagiarism of Saxton began in earnest after his 'privilege' came to an end in 1600. John Norden (1603) and William Smith (1602-3) sought in their abortive mapping projects to improve but not to supersede Saxton. Others, including the most well-known of them, John Speed (1611), were remarkable mainly for their slavish, albeit artistic, copies of the same material. John Speed's *Atlas,* the greatest of the derivatives, again went into nine editions, with some having several issues. Later, the Dutch map-makers Jansson and Bleau (1645) based their county maps on those of Saxton/Speed as did Robert Morden in 1695, and the armies on both sides of the Civil War used Saxton's maps. Even in the 1740s when Emmanuel Bowen and others engraved the county maps for *The Large English Atlas,* Saxton's topography is still discernible in most counties. The reason for this longevity was the absence of new field-work. Saxton's authority finally began to recede only as and when the counties were re-mapped after new original field-surveys had been undertaken.

The first to be released from the trammels of this cartographic bondage was Cornwall which in 1699 thus became the first county of England to be mapped on what was for the late seventeenth century the large scale of nearly one inch to the mile.[7] This remarkable pioneer venture by Joel Gascoyne made a significant contribution to English regional cartography, and the circumstances which led to this primacy for Cornwall require explanation. Joel Gascoyne was baptised on 31 October 1650 in Hull and subsequently 'placed himself as apprentice to John Thornton citizen and draper of London to serve a term from the day of the granting of the indenture for seven years', that is, from 21 October 1668. John Thornton was a platt-maker whose premises were 'at ye signe of the Platt in the Minories'.[8] He was one of a small group of platt-makers who plied their trade in shops lining the streets and alleys that led down to the waterfront on the north bank of the Thames down-river from the Tower of London. Throughout the seventeenth century these Thames-side manuscript chart-makers became linked in a master-apprentice relationship in the Drapers' Company. As an apprentice to John Thornton, Joel Gascoyne acquired the skills of chart-making, engraving and surveying.[9] This was a rare combination and one probably particular to John Thornton, for no other platt-maker of the Thames-side school is definitely known to have been an engraver and surveyor as well as a chart-maker.[10] Moreover, John Thornton produced much of the finest marine cartographic work done in England at the time.

In 1675, apprentice-days now behind him,[11] Gascoyne set up business at 'Ye Signe of ye Platt neare Wapping Old Stayres three doares below ye Chapell'.[12] Gascoyne remained a member of the Drapers' Company and paid his dues until the 9th of February 1689, when the last entry appears for him in that Company's Quarterage Book. At this point the whole direction of his professional life appears to have changed, and from 1690 to 1703 he seems mostly to have concerned himself with land-surveying. By 1693 it is clear that Joel Gascoyne had established himself as one of the leading land-surveyors of the day, working for such eminent individuals as John Evelyn, James, the third Earl of Salisbury, and their Majesties William and Mary.[13] At this point in his career he was to be wooed west from the capital city and from Government commissions. Gascoyne stayed in Cornwall for six years, during which period he produced the Stowe Atlas of thirty-three unsigned estate maps for the Grenville family. Like many estates of feudal origin, the land-holding here of the Grenvilles embraced a number of contiguous farms and covered a compact area extending over the parish of Kilkhampton. In complete locational contrast were the possessions held by the Robartes of Lanhydrock, who had accumulated their great fortune mainly by trade. This is reflected in the nature of their land-holding which was extensive in the whole but made up of pieces of land scattered over the county on which the successful merchant and dealer invested his accumulating wealth or made loans on mortgage, subsequently entering into possession on foreclosure. The mapping of the 258 properties which made up the Lanhydrock Atlas thus gave Gascoyne the opportunity of travelling extensively in the county. Between 1693 and 1699 these commissions enabled him to do not only the field-work for the survey of the manors but also the ground survey for his map of Cornwall (Figure 1.4).

The extension from the surveying of individual manors to the production of a topographical map of a whole shire by Gascoyne was, nevertheless, a big step, and would have been impossible without the active encouragement of the family for whom he worked. There is little doubt that Joel Gascoyne enjoyed the confidence and that essential and sustained support of a cartographically-minded patron and this combination of circumstances goes far to explain why Cornwall enjoys the distinction of being the first county to be surveyed on a large scale; it also accounts for the novel and sophisticated nature of some aspects of this map, such as the inclusion of parish boundaries.[14] For this it is necessary to enquire further into the nature of Gascoyne's patron; he was, as the cartouche on the map (Figure 1.5) makes clear 'Ye Rt. Honourable Charles Bodville Earl of Radnor, Viscount Bodmin, Lord Robartes, Baron of Truro Lord Lieutenant & Custos Rotulorum of Ye County of Cornwall'. By this time in Cornwall the original military function of the Lieutenancy and the civilian functions of the *Custos Rotulorum,* literally Keeper of the Rolls, had been merged; Robartes was therefore the chief officer in the county. Under his aegis would have come a whole range of military and civilian activities. Expressed in terms of space most of these were organised on a parish basis, hence the need to have this vital frame of reference expressed graphically, but with 201 parishes to oversee there was also the requirement to locate them efficiently. To facilitate this process Gascoyne filled the southern voids of his rectangular-framed map with lists, firstly of 201 parishes and then of some 2,475 places all of which were assigned to their respective parishes and in turn to the map-reference system. Furthermore, in distant and peripheral Cornwall in the 1690s it is difficult to envisage the compilation of such an extensive and comprehensive gazetteer without access being made readily available to official sources, presumably in the possession of the Lord Lieutenant.

Joel Gascoyne's association with the mapping of the West Country was not intended to end with the publication of the map of Cornwall, as in the following year he was the 'Undertaker' who issued the *Proposals* illustrated in Figure 1.6. Like many of Gascoyne's maps there is no name attached, but the first line leaves no doubt that he was the author of them and that he intended to produce a map of Devon closely similar to that 'already compleated' for Cornwall. Five hundred subscrip-

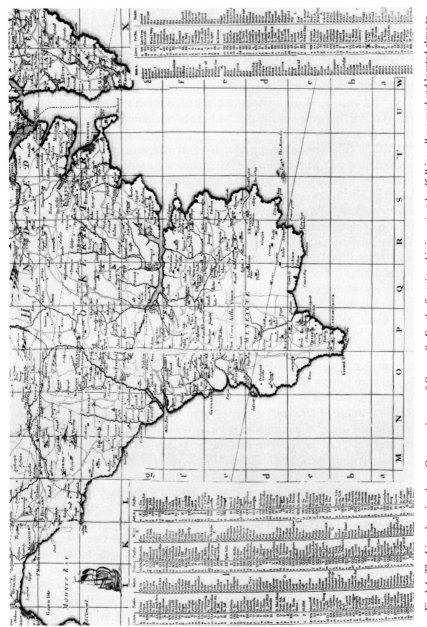

Fig.1.4   The Lizard peninsula on Gascoyne's map of Cornwall.   For the first time this important landfall is well mapped and located close to modern co-ordinate determinations.   Lizard Point was and remains the landfall for ships entering and leaving the English Channel.

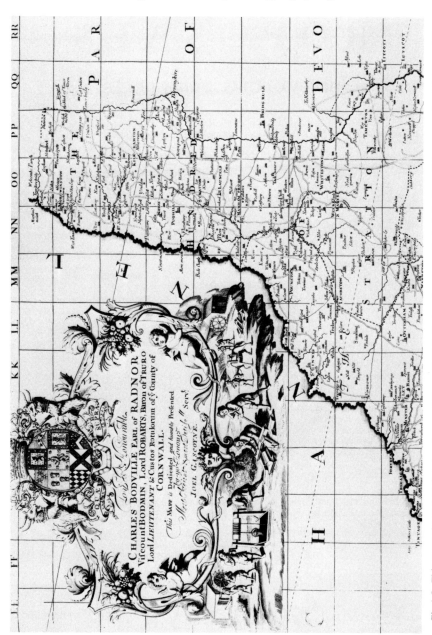

Fig.1.5 This cartouche on Joel Gascoyne's map of Cornwall 1699 displays the map-maker's dedication of his survey to Lord Robartes. The arms of the Robartes family are set above a floral effusion below which is a workaday scene of contemporary Cornish industrial activity.

PROPOSALS

Made for an
Actual Survey of the County of DEVON

THE *Undertaker* having already compleated a large MAP of the County of *Cornwal*, to the great Satisfaction of the Gentlemen of that County, is now defired by many Gentlemen of the County of *Devon*, to complete a MAP of that County : Whofe forward and earneftnefs therein, gives hopes to the *Undertaker* that others will follow their Example, to encourage fo generous, ufeful, and elaborate a Work. He therefore Propofes,

That if the Gentlemen of this County will Subfcribe for fuch a number as will anfwer the Charge and Trouble, he is willing to proceed in making a Map about fix Foot wide and five deep; wherein fhall be laid down their true Pofitions and exact Diftances, all the Hundreds, Towns, Villages, Gentlemen's Seats, principal Bartons, High Roads, Crofs Roads, Rivers, Bridges, and other Remarkable Places in this County.

This Map will be fo methodically Digefted and furnifhed with Tables, that any Place therein expreffed may be immediately found : And befides the Scale of Proportion in the Map, the Diftance of any two Places may be difcover'd at fight, without the help of Compaffes.

Every Map to be pafted upon Cloth, neatly Coloured, with Laquer'd Roles and Ledges, and Deliver'd to the Subfcribers at *Twenty Five Shillings* per Map.

Five Hundred Maps thus compleated, will but Reimburfe the *Undertaker*; and under fuch a Number it cannot be done.

And in confideration of the great Trouble and Expence the *Undertaker* will be at in Performing fo great a Task, he defires that every *Subfcriber* (for Encouragement) may pay *Ten Shillings* at Subfcribing, and the reft at the Delivery of the Map.

Such Gentlemen as defire to have their Coat of Arms *cut upon the Plate and Printed with the Map, are defired to fend them Blazoned to fuch places as the* Undertaker *fhall appoint in the feveral Parts of this County*; *and care fhall be taken (provided he may have as many as will make a Border) that they be well done, placing them Alphabetically*; *paying* Five Shillings *for the Engraving.*

Every Gentleman who Subfcribeth, fhall have his Name engraven under his Seat. Thofe who Subfcribe for *Seven* Maps, fhall have one *Gratis* ; and fo proportionably for a greater number.

The Perfons appointed to take SUBSCRITIONS, are,

In London, *William* Berry in *Craggs Court, Charing-Crofs.* At the *Grecian Coffeehoufe* in *Deveraux-Court*, near the *Temple*. *Chr. Brown at the Weft End of St.*Pauls *Church*. *John Thorneton* at the Sign of the *Plat* in the *Minories*. *Richard Mount*, Bookfeller, on *Tower-hill*.
Charles Teo in *Exon*. *Henry Chalkin* in *Taunton*. *Chriftopher Hunt* in *Barnftable*. *Tho. Michilfon* in *Bytheford*. *Hum. Burton* in *Tiverton*. *Anth. Paine* in *Torrington*. *Tho. Pollington* in *Newton-Abbot*. *Francis Hill* in *Plymouth*. *John Atherton* in *Totnefs*.

London, Printed in the year 1700.

Fig.1.6   Joel Gascoyne's *Proposals* for making a survey of the County of Devon 1700. Reprinted by permission of the Trustees of the British Library.

tions of twenty-five shillings, a total of £625, were required to 'Reimburse the Undertaker'. Gascoyne, in seeking such a large number of advance subscriptions, was being over-optimistic in terms of the early eighteenth century. Surely, he must have been hoping for substantial support from a few of the more wealthy and powerful 'Gentlemen ... Whose forward and earnestness therein, gives hopes to the Undertaker'. In the year 1700 Joel Gascoyne returned to London, where, as the *Proposals* make known, he set up the machinery for receiving subscriptions. Alas! in spite of all these good intentions, the necessary support did not materialise, and the survey of Devon was never completed, if, indeed, ever started. Although Gascoyne was a pioneer in his specifications for county mapping, his *Proposals* must be entered in the annals of cartography among those, and there were to be many subsequently, in which neither the 'noble, clergy and gentry' came forward with the necessary finance nor the market place with buoyant expectations.

To follow is always easier than to lead, nevertheless, it was some forty years later that the formidable challenge of mapping in the West Country was again accepted. This time it was by the Cornishman Thomas Martyn who parallels almost exactly Gascoyne's success in Cornwall and, regrettably, his failure in Devon. Like Gascoyne, he is known to have practised as an estate surveyor and a number of his manuscript maps have survived which bear dates in the 1730s. Around 1740 he embarked on the fieldwork for a map of Cornwall which, as this advertisement in *The Sherborne Mercury* for the middle of February 1748 shows, was 'now engraving, and will be published by subscription in about two Months Time'.

WHEREAS a new and accurate Map of
Cornwall, made from an actual Survey of the whole County, by
Thomas Martyn, is now engraving, and will be published by Subscription
in about two Months Time; The Nobility and Gentry that please to sub
scribe, are desired to send the Blazoning of their Arms before the
middle of March next, to Mr. John Richards, Mathematician in Exeter,
Mrs.Smithurst, Bookseller in Plymouth, Mr. William Rawling at South
Petherwin, near Launceston, Mr. James Jewell, Merchant in Truroe, Mr.
John Thomas in Marazion, Cornwall, or to the said Thomas Martyn at the
Golden Spectacles against Exeter Change in the Strand, London, where
Subscriptions are taken in.
PROPOSALS.
The Map will be six Feet long, and four Feet and half broad, and
printed on Imperial Paper.
The Arms of the Nobility and Gentry who are Subscribers to the Work,
will be engraved in the Margin.
The price to the Subscribers will be One Guinea; half to be paid at
Time of Subscribing, and the remainder on the Delivery of a Map in Sheets.

The map was eventually 'Published According to Act of Parliament November ye 10th 1748'.

The response to his *Proposals* must have been encouraging and so must have been the sales of the map of Cornwall during the first two years of its existence. One is prompted to draw this conclusion because two smaller-scale versions were published in 1749, and before the close of the year 1750, Martyn was prepared to embark on a similar venture to survey the large county of Devon. His *Proposals* appeared in *The Western Flying Post or Sherborne and Yeovil Mercury* on the 5th and 12th of November 1750.

PROPOSALS for surveying and making a new
MAP of Devonshire, by Subscription. By
Thomas Martyn, Author of the new Map of
Cornwall.
                    CONDITIONS.
    1. THE County shall be carefully survey'd, and Villages plotted,
by a Scale of an Inch to a Mile, by the said Thomas Martyn;
and the Roads, Rivers and Boundaries of Parishes, shall be described by
other Surveyors, that he will get to assist him.
    2. The Copper-Plates shall be engraved by the best Hands, and the
Names of Places engraved in Roman Character, (as in the large Map of
Cornwall) and the Names of the Nobility and Gentry prefixed to their
Country Seats in Italick.
    3. The Price to the Subscribers will be One Guinea in Sheets; one
half to be paid at the Time of subscribing, and the other on Delivery of
the Map.
    4. The Survey will begin in March next, provided there are a
sufficient Number of Subscribers;but if the Sum subscribed is not enough
to pay the Surveyors, then the Money shall be returned.
    Subscriptions are taken in by Mr. Thomas Blake, Gold-
smith, and the Booksellers in Exeter ... and by the said Thomas Martyn,
at Mr. Stephenson's, against Exeter Exchange, in the Strand, London.

From paragraph 4 his readers learned that the survey was planned to begin in March 1751, 'provided there are a sufficient Number of Subscribers'. However, a map of Devon from the hand of this Cornishman was not to be, for in the Ashburton Parish Register of Burials lies the entry '26 December 1751 Martyn, Thomas, a Stranger'. This tantalizingly cryptic statement poses many questions. If it can be assumed that sufficient support for the mapping of Devon had materialised, was he so engaged when his life came to an abrupt end in the parish of Ashburton? A further comment in a letter written on 25 January, 1752 by the Rev. Jeremiah Milles to the

Rev. William Borlase suggests that this was the case; it states 'Martin our map maker died at Ashburton last Xmass day, so that I fear we shall never get our county well laid down. He caught cold in traversing Dartmoor this wett summer, w[hi]ch ended in a milliary fever.'[15] True to his name, the plaint of Jeremiah Milles proved correct but only for the next thirteen years.

The significance of this mapping and attempted mapping of Cornwall and Devon can best be appreciated when they are placed in the context of what was happening in the rest of the country. It is now known that the second half of the eighteenth century was a crucial and distinctive period in the development of English regional cartography; a much improved county-based map coverage emerged through the enterprise of private cartographers. To assess the extent of this overall improvement, glances at the state of map-making at three sequential times in the eighteenth century are most revealing. If the distribution of completed county surveys at the one-inch to a mile scale or larger is taken as a rough yardstick of progress, then by 1750 only eight counties had been mapped. By 1775 a marked acceleration in the completion of surveys had occurred with something approaching a half of the counties of England being available. By 1800 only the western shires of Wales and the northern counties of Scotland had not been mapped at the one-inch scale or larger.[16]

This marked acceleration owed much to the far-sighted and energetic Cornishman, William Borlase. During the eighteenth century London coffee houses played an important rôle in the intellectual life of the capital and country. They functioned like embryo clubs, where for the price of a cup of coffee customers could foregather for conversation and discussion. Certain coffee-houses attracted particular clientèles. Rawthwell's at 25 Henrietta Street, attracted men of science, Fellows of the Royal Society and the like. On 22 March, 1754 certain 'Noblemen, Clergy, Gentlemen, and Merchants' foregathered and brought into being 'The Society for the Encouragement of Arts, Manufactures and Commerce', known since 1847 as The Royal Society of Arts. The aim of the society was to raise a fund to encourage and reward discoveries and inventions. The prime mover in getting the Society established was Henry Baker, F.R.S. A friend and fellow scientist was William Borlase, F.R.S., D.C.L., Rector of Ludgvan and Vicar of St Just in Cornwall. Though Borlase passed most of his life in this quiet corner of the Land's End peninsula, it was a life of unceasing energy in which by correspondence he advanced various scientific activities.[17] A letter from Baker to Borlase on 21 January, 1755 had contained the invitation 'if you can think of any Art, Manufacture or Improvement that by encouragement may become a public Good; if you will please let me know, I will lay it before them', that is members of the Society. Borlase's reply has become a basic document in the annals of British cartography:

> I would submit to you, as a friend, whether the state of British Geography be not very low, and at present wholly destitute of any public encouragement. Our maps of England and its counties are extremely defective...and the headlands of all our shores are at this time disputed...

Any improvement, he wrote was to 'be dispaired of, till the Government interposes, and attempts what would be so much for the honour as well as commerce of this island, 'tis to be wished that some people of weight would, when a proper opportunity offers, hint the necessity of such a survey'. He goes on to suggest that if the Society pump-primed by offering a reward for the

> best plan, measurement and actual survey of a city or district...it may move the attention of the public towards Geography, and in time, perhaps, incline the Administration to take this matter into their hands (as I am informed it does in some foreign countries), and employ proper persons every year from actual surveys to make accurate maps..., till the whole Island is regularly surveyed.

There is little doubt that Borlase was referring in particular to France where Louis XIV and his hard-working minister Jean Baptiste Colbert had much earlier initiated mapping by the state and had provided that essential leadership to ensure the continuing success of state-sponsored surveying.[18] Borlase, like the few people of vision in every generation, was decades ahead of his time. His clear articulation of the case for a government-financed national survey, although re-iterated as forcibly in December 1756, came to naught and the Society in March, 1759 decided to act independently and offered a 'premium for an actual survey of a county or counties'.

In October, 1759 Benjamin Donn of Bideford submitted his ideas to the Society and, having obtained their approval, published his *Proposals*. Seven years later and after several meetings of the Society, Donn became the first to receive a premium on 29 November, 1765 as is well-known locally since a facsimile was published some 26 years ago.[19] The Society's stimulus extended beyond those who actually won premiums. There were a number of rejected surveys and others were suggested but not completed. Some prize-winners prompted by their success went on to pursue their activities in other counties. Since the Centre for South-Western Historical Studies covers the four western counties it is appropriate to include here the fact that one of the rejected surveys was that of Dorset by Isaac Taylor; in truth the two county maps were being considered at the very same meeting of the Society. Taylor's Dorset was pipped at the post as it were on the apparent imperfections in its orthography of the place-names. In view of the contemporary variants in the spelling of place-names poor Isaac Taylor seems to have been given an unwarranted rough ride by the Society who seem to have been exercising extreme caution in awarding their first premium on the premise that it would be setting the standard for subsequent surveys.

Of the claims for awards in the fifty year period, that is between 1759 and 1809, only eighteen counties were successful.[20] One other county in the South West to secure this much coveted award was Somerset in 1782. By this date the Society had discontinued its larger premiums. Instead it gave 'honorary marks of...approbation' in the form of twenty guineas and a silver medal to William Day and 'the greater Silver Pallet to Charles Harcourt Masters'.[21]

Let us return to Devon, where most of what is new in the history of cartography for the latter part of the eighteenth century is associated either directly or indirectly with documents found at Ugbrooke Park. You will recall that Ugbrooke has been the seat for many generations of the Cliffords who rose to fame and distinction with the restoration of Charles II. It was the first Lord Clifford who entered history disguised somewhat under the initial letter of the acronym CABAL, and who was the architect and custodian of that memorable piece of duplicity the Secret Treaty of Dover. By it Charles II pledged to join Louis XIV in making war on the Dutch and, more importantly for this context, to declare himself a Catholic. The plan misfired and Charles wriggled out of his commitments. Not so Lord Clifford, he persevered, he did become a Catholic and the family have remained so.[22] The knock-on effect of that for cartography was considerable. But let us allow the documents to reveal their own secrets and to begin with, the rare, probably unique, survival of *Proposals* illustrated in Figure 1.7.

Owing to the undoubted success of Donn's map in 1765, it certainly came as a great surprise to me to find this broadsheet. It points clearly at another surveyor attempting to repeat the mapping of the county only some twenty years later, that is in 1787, when Donn's map was still available and he, himself, still active in cartography, but this is exactly what Richard Cowl proposed to do.[23] It prompts the question what was there sufficiently new and original that Richard Cowl had to offer in order to tempt a large number of the gentry to subscribe for yet another map of Devon? The *Proposals* promised a 'full, comprehensive and compleat' map of the county 'and its environs, viz. about one-third Part of Cornwall; some Part of Somerset; and some part of Dorset'. It would have depicted a larger area than that of Donn's map but since both maps were designed to be on twelve sheets one envisages Cowl filling in the voids right up to the neat lines of the large rectangle formed by the twelve copper plates. Apart from this, it is difficult to envisage what Cowl's map would have finally looked like since the *Proposals* are fairly typical of the promotion literature of the map trade in the last quarter of the eighteenth century. Note how the dictation of the market place persisted; the map would show what the 'Patronizers' would find 'valuable and entertaining'.

An intriguing new detail is the inclusion of 'hills and vales'. As the representation of relief was soon to become of prime importance one is left wondering whether Cowl was about to launch a new technique of portraying it. Most likely is it that he *xx maws 1787*

PROPOSALS,

For PUBLISHING by SUBSCRIPTION,

A FULL, COMPREHENSIVE, and COMPLEAT

MAP of the County of Devon,

And its ENVIRONS, viz.

About one-third Part of Cornwall; some Part of Somerset; and some Part of Dorset;

In TWELVE SHEETS,

(Which will form a Map, when joined, Seven Feet by Six nearly,)

Describing the Sea Coast, from Lime, in Dorset, to the Neighbourhood of Fowey, in Cornwall, being about an Hundred Miles; and from Bridgewater Bay, in Somersetshire, to the Westward of Bude, in Cornwall, being likewise about an Hundred Miles.—Also the Great and Bye-roads, the Bays, Harbors, Creeks, Ports, Fortifications, Castles, Forests, Ruins, Rivers, Brooks, Lakes, Hills, Vales, Hundreds, Parishes, Tythings, Hamlets, Churches, Chapels, Towns, Villages, Mansions, Farms, Antiqüities, and every Thing necessary to make the Work valuable and entertaining to the PATRONIZERS; from an accurate Survey performing

By R. COWL, of Plymouth,

And ASSISTANTS,

By whom Lands are accurately Survey'd, and neatly Mapp'd and Embellish'd.

N.B. The SHEETS to be delivered PLAIN.

CONDITIONS.

I.

THE Work to be well Engraved, by the most capital Hands, and Printed on good Paper.

II.

The Price to Subscribers will be Three Guineas; One-half to be paid at the Time of Subscribing, and the Remainder when the Work is compleated, which will be forwarded with as much Expedition as the Nature of the Business will admit of.—The Price to Non-subscribers will be Four Guineas.

1787. March the 2nd Received of N. Thom. Hugh Clifford Lord the Sum of one guinea half being the first Payment of Subscription for the County of Devon, ₤.11.6 ℟ Rich. Cowl

Plymouth: Printed by R. Trewman and E. Haydon, Letter-press and Copper-plate Printers.

Fig.1.7   Richard Cowl's *Proposals* for making a new 'Map of the County of Devon' in 1787.

intended to do something similar to what appeared on his 1780 map of the Plymouth area, where he puffed up the river systems with spurious valley forms.[24] Another unusual feature, although not unique, is the inclusion at the foot of the page of a means whereby a receipt for the subscription might be entered, in this case the 'One half to be paid at the time of subscribing'. Such a method of advance financing of long-term pieces of work of this kind had distinct advantages as far as the cash-flow problems these surveyors usually had, but it was an advantageous financial arrangement only if all the pieces of the map-making jig-saw fitted satisfactorily together. Sadly, for Richard Cowl they did not as the *Exeter Flying Post* poignantly reported on 20 August, 1789, 'Thursday last Mr. Cowl, of Plymouth, who had surveyed, and was about to publish a Map of the County, had the misfortune to be thrown from his horse, near Bickleigh, on his road from Tiverton to Exeter, and killed on the spot'. The survey was nowhere near so advanced as the report in the paper made out and it never was finished.

A document such as that speaks mainly for itself. Not so that which is illustrated in Figure 1.8; just one of a collection of seventy-seven maps whose cartographic content is meagre consisting as it does of the coastal outline with streamlines in estuaries, rivers and small circles indicating the major settlements. Save for a dotted line to demarcate the county boundary there are no other topographical, cultural or textual details. With so little information and at first no supporting documents, the process of discovering their purpose and ascription to authorship presented quite a problem.[25] Water-marks in the paper ranging from 1794-1804, a scale of 5 miles to the inch and other clues eventually showed these to be outlines copied from John Cary's *New Map of England and Wales* first published in 1792. Their true purpose, however, did not emerge until it was fully appreciated that the cartographic archive at Ugbrooke stemmed principally from the Honourable Robert Edward Clifford (1767-1817), who emerges as one of the pre-eminent military analysts and cartographers of the period (Figure 1.9), and who became closely associated in a consultancy capacity with John Cary the head of one of the most prolific map-publishing houses in London.[26]

As a consequence of the Penal Laws it was quite normal for Catholic families to send their sons abroad to be educated and brought up in the Faith. In 1776 Robert Edward entered the Academy for English Catholic youths at Liège.[27] Commissions in the British Army at this time were not open to Catholics so Robert Edward received one from Louis XV1 and served in Dillon's regiment of Irish Jacobites.[28] Then he states, 'I gave my dismission from the french service in 1791 stating conscientious motives for not following the french revolution';[29] not surprising in view of the increasingly anti-catholic stance the Revolution was adopting. Here then was an officer who for years had been trained in a crack French regiment and who at first hand knew the ways in which the French forces were instructed and operated. Such

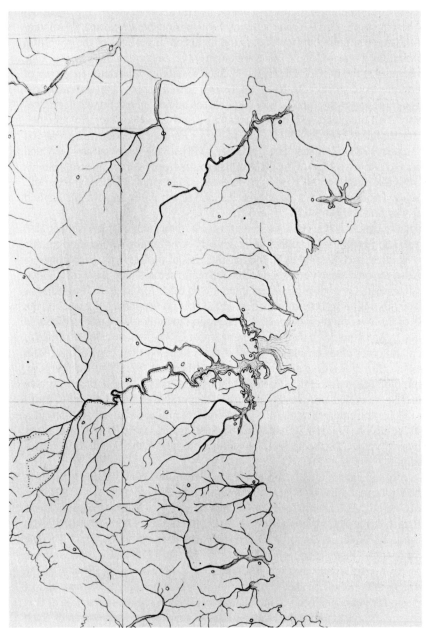

Fig.1.8 A reduced reproduction of the 'skeleton' covering parts of Devon and Cornwall. Note the limited amount of information on the map so as to provide as much uncluttered space as possible for officers to fill in details of importance for their military needs. The end product was the so-called coup d'oeil frequently made by sketching in features while on reconnaissance marches.

Fig.1.9   A photographic copy of the portrait of the Honourable Robert Edward Clifford painted by Mather Brown. It hangs in the entrance hall stairway at Ugbrooke and shows Robert Edward with a map in his right hand.

an officer was an invaluable source of intelligence to the British as the Napoleonic conflict intensified, particularly when it is remembered that throughout the eighteenth century the principal centres of innovation in military science were located on the continent. By the end of the century it was the French army's system of organisation, its military thinking and provision for officer education which had reached the apex of Europe's military culture. To the long-established strategies of erecting and subduing fortified strongholds, and the conducting of set-piece battles was added the new and different war of movement; henceforth, armies were well-constituted, mobile and able to manoeuvre. A direct consequence of this was the further enhancement of the rôle to be played by maps; both the making and acquiring of them became of increasing importance. It was again the French who had made the greatest strides in the application of cartography to the logistics of an army on the march and who, in any case, had now assumed the mantle of European leadership in the realm of scientific map-making.[30]

Robert Edward left France imbued with this up-to-date knowledge of military affairs to find a number, at least, of like-minded, forward-looking, innovative, adaptive, military minds in England. Foremost among these, as far as our context is concerned, was John Graves Simcoe.[31] He had served in the American Revolutionary War and between 1792-96 was the first Lieutenant-Governor of the Province of Upper Canada. At the end of this tour of duty and another in St. Domingo he returned to his home at Wolford in east Devon where he renewed his friendly relationship with the Clifford family. In 1798 when a general mobilisation was ordered in Britain to prepare to repel any invasion from across the Channel, General Simcoe was given the command of the Western District, that is Cornwall, Devon, Dorset and Somerset. One of his immediate needs was to provide himself with maps. None suitable was available. In such a time of great need Simcoe turned to Robert Edward who set about producing a set of 'skeleton' military maps covering the whole of southern England as far north as a line joining Anglesey to the Wash. The production, which he made clear in a letter surviving in the Simcoe archive, clearly followed Continental practice, and that was:

> to take the course of all the rivers laid down in Carys large map of England, and the position of all the towns in Capital letters; This when taken upon oil paper, I engrave on large plates of copper, so that we may have as many skeletons of England as we please. Not a word of writing will be seen; that is to be done with the pen, and nothing to be written with the pen but what has immediate relation to the author read. It looks like the anatomy of the Veins, and the towns so many o 's; Having Cary's great map it is very easy to fill up just what one pleases... This would be of great use for officers going to the outposts, as in 5 minutes they may take the position of all the roads & passes within three miles of their post from the general map & keep it in their orderly book. This would form officers to understand the value of positions, & give them a desire of taking &

drawing plans, hence they would acquire that coup d'oeil which blind commanders seldom acquire,...[32]

Well! maybe all right for a person of Robert Edward's ability and training but probably a great deal to ask from the ordinary officer!

The need for this cartographic expedient was dictated by the sorry fact that no suitable military maps (official or otherwise) were yet available. Let us look back a few years. The spatial and temporal context was that some of the cognoscenti in late eighteenth-century England had been stung into action by the French proposing in 1783 that the long-standing and at times acrimonious disagreement over the longitude difference between the Paris and Greenwich observatories be finally settled by the combined efforts of English and French cartographers.[33] Standing in the wings, having waited many years for such an opportunity to undertake the initial basic triangulation as a preliminary to a national survey, was Lieutenant Colonel William Roy. A venture as prestigious and as politically sensitive as this called forth the combined efforts of the Board of Ordnance, the Royal Society, and not least and at long last a royal lover of maps. George III not only approved the plan and the finance but also graced the work of survey with his presence. Thus came into existence the institutional structure to provide maps for the nation but with the needs of the military well to the fore. The trigonometrical link across the Straits of Dover was completed on 17 October 1787; it was to be the last major cartographic activity undertaken by the ancien régime.

Two years later the tocsin of Notre Dame summoned the mobs to the streets of Paris. In 1793, when hostilities with France formally began, priority in mapping by the fledgling Ordnance Survey was given to those counties bordering the English Channel, the invasion coasts so called. The triangulation reached Land's End in 1796, much earlier than would otherwise have been the case. The topographical mapping, as distinct from the trigonometrical survey, was underway in Devon in 1803, the year in which the fragile peace of Amiens came to an end and an invasion by Napoleon was being planned. Lt. Colonel William Mudge, a Plymouth man, and the first effective director of the Ordnance Survey, set up his temporary headquarters at Chudleigh, the nearest small town to Ugbrooke.[34] Contacts between Mudge, Lord Clifford, who was chairman of the committee for the internal defence of Devon, and General Simcoe were soon established with Mudge providing copies of the Surveyors' Drawings as they became available. These were manuscript maps, on a scale of three inches to the mile, emphasising relief, drainage, vegetation cover, roads, that is those aspects of the terrain vital to troop movements and concealment. Two copies, coloured to show the land-holding of the Cliffords, are extant at Ugbrooke and must be very rare examples surviving outside the official corpus now lodged in the British Library. When reduced, the Surveyors' Drawings were compiled and edited to make the one-inch 'Old Series' Ordnance maps, those of Devon eventually appearing for sale in 1809.

It may be thought that this would strike the death knell for the private map-makers who had functioned without official competition since the time of Queen Elizabeth. The likes of John Cary were not prepared to go down without a fight. Early pulls from copper plates, an example of which is illustrated in Figure 1.10, were sent to Robert Edward Clifford presumably for his comments and appraisal; they contain enough detail to detect a clear copying with a half-scale reduction of the Ordnance Survey sheets. The mystery of their existence deepens when it is realised that in 1811 the Master General of the Ordnance prohibited as a military precaution the sale to the public of all Ordnance Survey maps, wisely in view of Napoleon's known voracious appetite for seeking out and hoarding such publications.[35] Nevertheless, in 1813, right in the very middle of these eventful years and with the ban still on, John Cary began to include in his sales catalogue a half-inch map of Devon. Since after much searching no such map seemed to exist and when almost ready to conclude this was just another anticipated but unrealised map-vendor's advertisement, deeper search beyond the library catalogues unearthed a copy (Figure 1.11). A close inspection reveals that in terms of national security and military intelligence these maps show most of the essential details of their larger originals. How Cary justified this seemingly unpatriotic action is bewildering. Moreover, you may well be wondering: what about copyright? Cary, with other London map-sellers, got round that problem by arguing that 'as a portion of the Public, at whose expense the Ordnance Survey is carried on, they had a right to reduce from and publish, Copies of the Ordnance Survey on Scales suited to their own convenience.'[36] Certainly, Cary was not secretive about his intentions as the title makes clear.

Is it then the only surviving copy? It is, but only to the best of my knowledge. Note my gloss, for this map joins, at a conservative estimate, the other six million maps either not catalogued or inadequately catalogued in the Public Record Office alone. By this recent disclosure[37] one must conclude that our national cartographic heritage remains, in the words of the medieval cartographers, *terra incognita* awaiting 'discovery'. I hope that this exegesis of the twenty-first Harte lecture 1990 will 'discover' for all its readers some new landmarks in the South West's cartographic heritage.

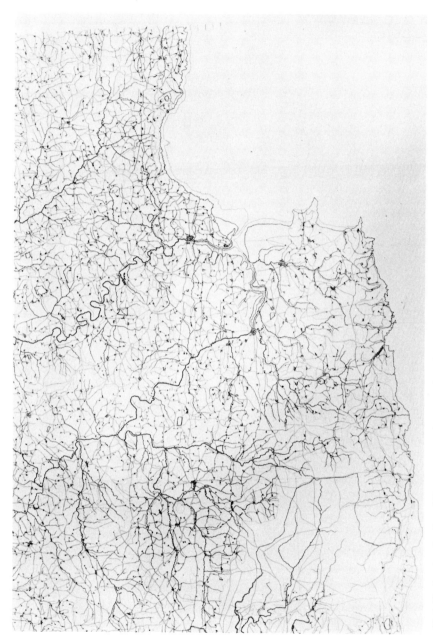

Fig.1.10 A reduced early pull from a copper-plate engraving of north Devon. This is a half-scale reduction of parts of Sheets XXVI and XXVII of the Old Series Ordnance Survey Map and was doubtless sent by John Cary to the Honourable Robert Edward Clifford for his critical appraisal.

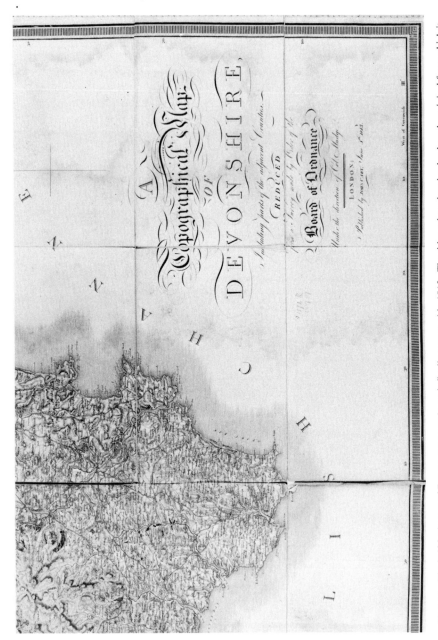

Fig.1.11 The half-inch map of Devon, not to scale, as it finally appeared in 1813. The title makes abundantly clear the original from which it was reduced.

Acknowledgements

The author is much indebted to Captain the Lord Clifford for his permission to consult the archives at Ugbrooke, for allowing Figures 1.7, 1.8, 1.9, 1.10 to be reproduced and for guidance and information concerning former members of the Clifford family.

Notes and References

1. R.A. Skelton, *Saxton's Survey of England and Wales with a Facsimile of Saxon's Wall-Map of* 1583, Imago Mundi Supplement No. VI (Amste dam,1974).
2. Sir Roy Strong, *The Portraits of Queen Elizabeth 1* (1987).
3. J. Arnold, *Queen Elizabeth's Wardrobe Unlocked* (1988).
4. R. Helgerson 'The Land Speaks: cartography, chorography, and subversion in-Renaissance England', *Representations,* 16 (1986), 50-85;   V. Morgan, 'The Cartographic Image of "The Country" in Early Modern England', *Transactions Royal Historical Society,* 5th Series 29 (1979), 129-54 and 'The Lasting Image of the Elizabethan Era', *Geographical Magazine,* 52, no.6, (March 1980), 401-8.
5. W. Ravenhill, *John Norden's Manuscript Maps of Cornwall and its Nine Hundreds* (Exeter, 1972).
6. R.A. Skelton, *County Atlases of the British Isles 1579-1703* (1970).
7. W. Ravenhill, 'Joel Gascoyne, A Pioneer of Large-Scale County Mapping', *Imago Mundi,* 26 (1972), 60-70.
8. Register of Holy Trinity Church, Kingston upon Hull, vol.1, 1553-1653; Drapers' Company Records, Bindings Book 1655-1689, +290; T. Campbell, 'The Drapers' Company and its School of Seventeenth Century Chartmakers' in H. Wallis and S. Tyacke (eds.), *My Head is a Map* (1973), 81-106; T. R. Smith, 'Manuscript and Printed Sea Charts in Seventeenth-Century London: the case of the Thames School' in N.J.W. Thrower (ed.) *The Compleat Plattmaker* (1978), 45-100.
9. C. Verner, Introduction to *The English Pilot The Fourth Book London 1689* (Theatrum Orbis Terrarum, Amsterdam,1967), vii; W. Ravenhill, 'The Marine Cartography of Devon in the Context of South-West England' in D.J. Starkey (ed.), *Devon's Coastline & Coastal Waters,* Exeter Maritime Studies, 3 (Exeter, 1988), plate VI and 12-14.

10. The evidence that Thornton practised estate surveying is not extensive but there is a map signed by him in the Bodleian Library, Oxford Maps Eng.a 2. I am indebted to Tony Campbell for this information.

11. Drapers' Company Records, Stamped Freedoms Book 1665-1746,+280.

12. This address, in various forms, appears on Gascoyne's charts of the pre-1690 period.

13. W. Ravenhill, 'Joel Gascoyne and the Mapping of Sayes Court in the Parish of Deptford, 1692', *Guildhall Studies in London History* 1, no.4, (1975), 250-2.

14. W. Ravenhill, 'The Making and Mapping of the Parish: the Cornish experience', in W.K.D. Davies (ed.), *Human Geography from Wales: Proceedings of the E.G. Bowen Memorial Conference, Cambria,* 12, nos. 1 and 2 (1985), 55-73.

15. Borlase MSS Original Letters 3. 12a. The author is indebted to P.A.S. Pool, Hon. Archivist, Penzance Library, for this reference.

16. P. Laxton, 'The Geodetic and Topographical Evaluation of English County Maps, 1740-1840', *Cartographic Journal,* 13 (1969), 37-54; J.B. Harley, 'The Re-Mapping of England, 1750-1800', *Imago Mundi,* 19 (1965), 56-67.

17. P.A.S. Pool, *William Borlase* (Royal Institution of Cornwall, 1986).

18. J. Konvitz, *Cartography in France 1660-1848* (Chicago, 1985).

19. *Benjamin Donn: A Map of the County of Devon 1765*, Reprinted in Facsimile with an Introduction by W.L.D. Ravenhill (Exeter, 1965).

20. J.B. Harley, 'The Society of Arts and the Surveys of English Counties 1759-1809', *Journal of the Royal Society of Arts,* 12 (1963-4), 43-6, 119-24, 269-75, 538-43.

21. *Somerset Maps, Day & Masters 1782 Greenwood 1822,* Reprinted in Facsimile with an Introduction by J.B. Harley and R.W. Dunning, Somerset Record Society, 76 (1981).

22. H. Clifford, *The House of Clifford* (1987).

23. W. Ravenhill, 'Richard Cowl's Proposals for Making a New County Map of Devon in 1787' *Devon & Cornwall Notes & Queries,* XXXV, Part IX (1986), 338-44.

24. *A Plan of the Town, Citadel, Dock, and Sound of Plymouth, with Their Environs,* Surveyed by Richard Cowl and Engraved by William Faden 1780, British Library, K.11.83.

25. W. Ravenhill, '"Skeletons" at Ugbrooke Park, Devon', *Cartographic Journal,* 21 no.1, (June 1988), 50-8. The conjectures about the maps of southern England being part of the production process must now be modified in the light of Robert Edward Clifford's (hereafter REC) letter to Simcoe of 1801; see below note 32.

26. Sir Herbert G. Fordham, *John Cary Engraver, Map Chart and Print Seller and Globe-Maker 1754 to 1835* (1925, reprinted 1976).

27. R. T. Lomax, 'Boys at Liège Academy 1773-91', *Catholic Record Society,* 13 Miscellanea VIII (1913), 202; Henry Foley, *Records of the English Province of the Society of Jesus* (1882), VII, part I, liii.

28. J.G. Sims, 'The Irish on the Continent 1691-1800', in T.W. Moody and W.E. Vaughan (eds.), *A New History of Ireland* (Oxford,1986), IV, 629-53; J. C. O'Calleghan, *History of the Irish Brigades in the Service of France* (Glasgow,1870); Capitaine Malaguti, *Historique du 87e Regiment d'Infanterie de Ligne 1690-1892* (1890), 77.

29. Ugbrooke Park Archives (hereafter Ugbrooke), General Correspondence (hereafter GC) 1808-15, Letter 27 Nov. 1814, REC to Bishop Poynter.

30. F. B. Artz, *The Development of Technical Education in France 1500-1850* (Cambridge, Mass. and London, 1966), 87-101; S. Wilkinson, *The French Army before Napoleon* (1915); J.B. Harley, B. B. Petchenik, and L. W. Towner, *Mapping the American Revolutionary War* (Chicago, 1978); R.H. Thoumine, *Scientific Soldier: A Life of General Le Marchant 1766-1812* (1968).

31. M. van Steen, *General Simcoe and his Lady* (Toronto, 1968).

32. John Graves Simcoe Papers, Archives of Ontario, Series A-4-1 Loose Correspondence (hereafter Simcoe Papers) Letter 1801 REC to Simcoe; J.G. Tielke, *The Field Engineer,* translated by E. Hewgill, 2 vols (1789), I, 5-6, II, 195; Y. Hodson, 'The Military Influence on the Official Mapping of Britain in the Eighteenth Century', *IMCOS Journal* 27 (1987), 21-30.

33. W.A. Seymour, *A History of the Ordnance Survey* (Folkestone, 1980), 12-18.

34. Simcoe Papers, Letter 18 May 1803, Mudge to Simcoe.

35. Le Colonel Berthaut, *Les Ingénieurs Géographes Militaires 1624-1831* (Imprimerie du Service Géographique, 1902), Tome 1, 278-85.

36. P.R.O. W.O.44/299 28 September 1816.

37. Public Record Office, *Readers' Bulletin,* 6, 1990.

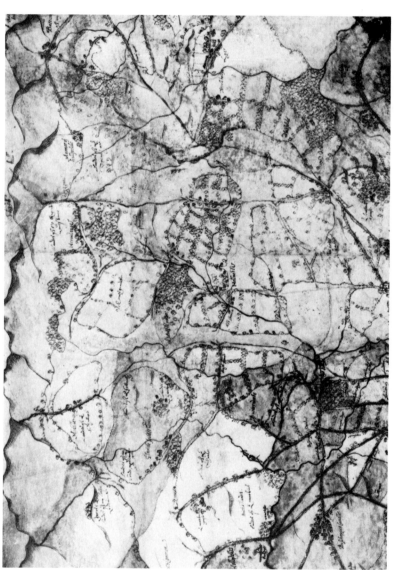

Fig. 2.1 Photograph of part of an Elizabethan map of north-west Dorset 1569-74 (British Library Add MSS 52252), showing an area around Holnest (cf. Figures 2.4, 2.5 and 2.6). Along the top (south) edge, the chalk scarp of central Dorset is represented: *Stoys Hill* (High Stoy) is labelled. To the right (west) is the village of *Lye* (Leigh). See Barker *op.cit.* plate 7 for comparison of the 1569-74 settlement plan with the six-inch Ordnance Survey 1888. Leigh enclosure is mentioned by Jennifer Baker, see Figures 5.12 and 5.13. Figure 5.12 shows the lane leading north-west from the church to Yetminster (until 1849 the route taken by burial parties), the enclosures are not shown 1569-74. The fields of Figure 5.13 date from the Leigh Inclosure Award of 1804. Photograph reproduced by permission of the Trustees of the British Library.

CHAPTER TWO

An Elizabethan Map of North-West Dorset: Sherborne, Yetminster and Surrounding Manors

Katherine Barker

In 1965 P.D.A. Harvey published a short paper on an Elizabethan map of the manors of north Dorset (BL Add MSS 52522) recently acquired by the then British Museum (Figures 2.1 and 2.2).[1] He gave a brief account of what little is known of the history of this map, followed by a description of its appearance, scale and style. He drew particular attention to the division of the area into manors, each distinguished by overall colouring. The name of each manor is given in an italic hand, and beneath, in a secretary hand, there is the name of the lord of the manor. From this assemblage of names, Harvey dated the map to between 1569 and 1574.[2] The map has no title, no key, and a small rectangular cartouche at both top left- and top right-hand corners have been left blank save for two groups of lightly written numbers; a third group is found against the right-hand margin (Figure 2.2). The map is without decorative embellishment.

Harvey noted that the survey corresponds to no known administrative area, and that its purpose was unknown. He suggested the map was commissioned by the owner of Sherborne Castle, the Bishop of Salisbury, who recovered it from the Crown by Bill of Chancery between 1556 and 1558. For while the rectangular format of this map will naturally preclude any close coincidence with an actual estate, the content and arrangement of the material is most easily seen - and indeed makes most sense - as the graphic expression of the territorial interests of a single individual for whom the mapmaker gives no name. The purpose of this paper is to add substance to the suggestion that this individual was indeed the Bishop of Salisbury, and further, that the content of the survey is likely to reflect the terms of the legal action in Chancery of a few years earlier.

Despite the rich topographical detail, Harvey noted that the map itself was based on an inadequate survey. The mapmaker has certainly treated the area unevenly,

29

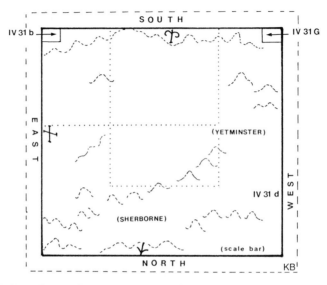

Fig.2.2   Scale diagram of the Dorset map format (actual size 84 cms x 77 cms approx);
compass references are written in a black border. Markings (enlarged) against the margin
may be correction symbols (compare Figure 2.3), figures could suggest continuation sheets.
Hills are represented by 'molehills' as indicated, and the dotted lines show the approximate
areas of Figures 2.1, 2.4 and 2.6.

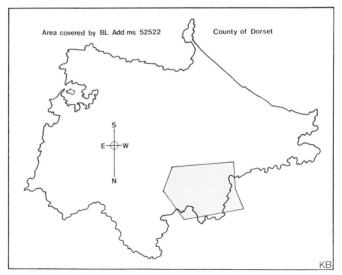

Fig.2.3   Dorset: stippling shows the area covered by the Elizabethan map, BL Add MSS
52522, dated 1569-74

some parts are very detailed and others are almost blank. There is considerable distortion, some areas are very compressed while others have been expanded out of all proportion with their neighbours. Yet despite this fact, the mapmaker has seemingly aspired to make it a scale drawing and he included a simple, unadorned scale-bar at the bottom right-hand corner; a *myle* reckoned in terms of a thousand *passus*, paces. The computation that accompanies it is in Latin. This internal information gives a total east-west measurement of approximately six miles, and a north-south measurement of some five and a half. In statute miles this would be rather in excess of nine miles by eight. Such figures do not, however, give the total area covered by the mapmaker which, as can be seen from Figure 2.3, is not confined within a regular border. The north and south edges are not parallel, the western edge is adjusted to include both Yeovil and Stoford, and the eastern edge is much deflected to make room for Lydlinch, an outlying part of the Sherborne hundred, although there was only just room for the village - the eastern part of the manor is off the map.

The map may have been unfinished; marks against the margin may indicate corrections to be worked into a later draft, and small groups of figures on the edges could hint of continuation sheets (Figure 2.2). Closer inspection however, leaves us in little doubt that the map was drawn from information gathered during an actual field survey which, inadequate as it is by later standards, may well have fulfilled its purpose quite satisfactorily. Whatever the apparent shortcomings, this is an Elizabethan document which deserves close attention. No written source from this period would be expected to yield, without effort, its content, meaning and possible historical significance. But this document is attractive, its immediate appeal is to the senses rather than the intellect.

Not only is the map highly coloured, it is richly pictorial; the landscape is seen from a bird's-eye view and 'reading' from the north-east - the bottom left hand corner. The reader is beguiled behind his lens; effortlessly his eye is led over the turrets of the castle, along the Tudor streets of Sherborne between the narrow-gabled houses, past the dovecotes, over the bridge by the mill, its wheel dipped in the blue water momentarily stilled, and southwards up the hill past the gallows and out among the green fields and hedgerows of the neighbouring villages. These scenes can be reconstructed from present experience. While the houses are 'placeless' (red roofs are not characteristic of the district as Harvey noted), some churches present individual features that can only have been drawn from life, and so may the water-mills and the wayside crosses. At this level the map is a picture, a powerful and immediate evocation of the landscape of today viewed from the past. Much less immediate is the appreciation of the mapmaker's meaning and purpose; reconstruction of the historic landscape is imposed from the vantage point of the here and now, but the mapmaker can only speak from his own time. The elucidation of his code and an understanding of the demands made upon him remain a continuing task, still in progress. And because this map is an archive with such a strong visual and geographical content, the challenges of transcription and interpretation are perhaps greater.

Correlation of the 'then and now' carries its own innate satisfaction; the 'matching' of church for church, mill for mill, and even hedgerow for hedgerow, the tracing of lost roads and hamlets, and the location of forgotten names and places, are all worthy pursuits. At this level however, omissions and inconsistencies severely limit the creation of a 'complete' picture; towards the centre of the map the information is very full, but towards the edges the information is disappointingly slight and to both north and south the survey 'tails off' into undifferentiated rows of 'molehills' representing, respectively, the Jurassic scarp that borders the Somerset Plain and the Chalk scarp of central Dorset, in which the mapmaker evinces little interest. The map is however, more than the sum of its missing parts. Omission is the natural complement of selection, and thus approached, this archive begins to betray a certain overall cohesion that speaks of the authentic product of a single purpose.

In the presentation of what frequently appears to be an almost random distribution of facts and figures, it will be argued that there lies a code and an intention which transcend what the twentieth-century eye too easily confines to the realms of the curious and picturesque. Whatever the technical shortcomings of the field survey - and we may only judge these from our own situation - the map is an important historical source, not only of the landscape, but of its own time. It dates from a decade which witnessed a rapid (and little understood) increase in the use and appreciation of maps among the educated and landowning classes of England, although not until the 1580s does it seem to have been generally understood that an estate survey might involve the making of a scale-map.[3] In an age of high inflation where social and political uncertainties were inseparable from the rapid turnover of land, the map or *Platt* (an expensive production) was for purposes of both litigation and prestige. As the former it was already in use in conjunction with the written survey, as the latter it was increasingly the symbol of territory and authority.[4]

For several centuries, authority in Sherborne had rested with the Bishops of Salisbury. The area surveyed for the making of the map contains the two episcopal hundreds of Sherborne and Yetminster, two of the three hundreds probably held by the Bishops of Sherborne in the early eleventh century.[5] The see was removed to Sarum shortly after the Norman Conquest, and it was an early twelfth-century Bishop, Roger of Caen, justiciar to Henry I, who built Sherborne Castle a little east of the town and alongside laid out a deer park.[6] Thus large-scale in its coverage, almost parochial in its attention to detail, the content of this map is nevertheless one which mirrors the temporal and spiritual pre-occupations of a kingdom at a time of unprecedented change.

Landscape history is a subject of wide confines and broad application. The intentions of the mapmaker were those defined for him; his proper purpose was the fulfillment of an aristocratic commission. His 'landscape history' was that necessary to record the *status quo* with a view to its future preservation. In the depiction of the two hundreds, the mapmaker has drawn a pair of territories each of which was bound not only within traditional patterns of lordship, but by complementary agricultural and economic ties that were the development of many centuries. A notable feature of this area is the well defined topographical 'banding' displayed, perhaps unwittingly, on the map. To the north, occupying the bottom third of the survey, is the wide limestone valley of the River Yeo (formerly *Yeovil*), an early Anglo-Saxon royal manor and site of the Sherborne bishopric from the early years of the eighth century. To the south, on the dip-slope, are a group of dependent settlements first recorded post-conquest, and occupying the top third of the map is the extensive open common and woodpasture of Blackmoor. Running across the map from bottom to top are a number of routes which linked the primary episcopal manor both with its 'daughter' minster, and with its outlying possessions, in a powerful statement of ancient rights and affinities which, in themselves, provide the survey with an underlying cohesion and validity. If the contemporary world has recourse to geographic determinants in the shape of multiple estate theory[7] then the Tudor world would perhaps have seen the identity of the two hundreds more in terms of a natural order of correspondence endowed and sanctified by God.[8] Osmund, eleventh-century (and first) Bishop of Salisbury, was known to have put a curse on anyone who alienated the manor of Sherborne from the bishopric.[9] If such a tale were already current in the mid-sixteenth century, then it gathered strength in the early years of the next century following the untimely ends of Henry, Prince of Wales, Sir Walter Raleigh and the Earl of Somerset.[10]

But the past was the mapmaker's pre-occupation only in so far as it concerned the placement of lordship in a contemporary landscape. Shades of green, turquoise, blue, pink, russet, ochre and yellow give each manor an identity. Their shapes are, at first glance, haphazard, a coloured mosaic dominated on the right hand by the large sub-circle of the Yetminster manor, second only to the great pale-green-hued expanse of Sherborne which fills much of the lower third of the map. The colours are arresting but can literally blur the bounds of these lordships to which, by definition, the colour-coding draws the attention. As clear are the differing patterns of enclosure which draw the eye to the central area of the map and which, in a number of places, are not coincident with the manors. A clear interest in fields - or at least in differing patterns of enclosure - will surely reflect an interest by the mapmaker in manor boundaries. Looking first at the manor of Holnest, this paper traces some of these manorial boundaries, not only in their association with field systems, but with the apparently unsystematic, almost random distribution of names and features these manorial boundaries, not only in their association with field systems, but with the apparently unsystematic, almost random distribution of names and features recorded on the map. Further, there is a distinct correlation between those manors for which no landowner's name is given, and those estates held by the Bishop of Salisbury from 1556 to 1558.

Over 170 place and minor names appear on the map, principal among which are those of manors. Some estates are not described as manors, and many are followed by the words *Landes of* ... and then a personal name, among which figure Strangways, Trenchard and Poulet, members of well-known West Country families. Among those of the 'lesser gentry' are Horsey (Clifton Maybank and Wyke), Mollins (Folke), Lewston (Leweston) and Thornhill (Allweston), whose escutcheons, together with those of John Jewel, Bishop of Salisbury (1559-71) were mounted on the south wall of the headmaster's house as principal benefactors to the King Edward VI grammar school of Sherborne founded in 1550. The house itself, now incorporated into the Lady Chapel of Sherborne Abbey, was built 1560-61; the arms have recently been restored. [11]

The distribution of other names is very uneven, some in italic and others in secretary hand. Over a dozen hills are named, including *Crackmacke* Hill (Crackmore, near Milborne Port) which appears to be topped by a cairn, and *bubdowne* Hill (Melbury Hill) the summit of which is marked by a beacon represented by a curious little device which bears little resemblance to the cresset and ladder arrangements which figure so prominently on a map of the Dorset coast dated to 1539. [12] Watercourses are a prominent feature on the north-west Dorset map, but only two are named. *Mark lake* (OE *mearc*, 'a boundary') is a small stream which flows along the boundary of Holnest manor, here probably also the Sherborne hundred boundary. The OE *lacu*, 'watercourse, a stream' is well evidenced in pre-conquest charters, and occurs again in *remslake* Bridge (*Rimeslake*, 1614) on the Bishop's Caundle/Down-Holwell boundary. The first element has been given as uncertain but once again a boundary could be inferred, this time in OE *rima* (Figure 2.4). [13]

The map shows many roads but names only three *lanes*, one *way* and the principal streets of Sherborne town. In a rural context, *street* describes a settlement of linear plan arranged along a routeway. Although there are several in the area, those shown by the mapmaker are all in Holnest, and they are *Weekestreete*, *Hethfelde streete* and *Est streete* (Figure 2.5). None is in current use. Of woods shown, only five are named, Oborne, Honeycomb (Sherborne), Clifton (Maybank), Holnest, and Holt (Stourton Caundle); a large wood is shown beside the name *Whitfelde*, and this was certainly Whitfield Wood, but others, including woods of the royal manor of Hermitage, are left nameless. A number of single trees are shown, and three are named; *A thorne*, *weeke ash* (both in the manor of Wyke), and *quckowe oake* in Holnest.

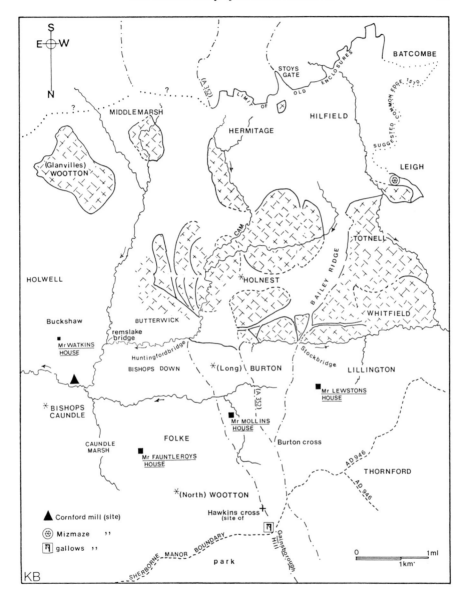

Fig.2.4 Area south of Sherborne: cross-hatching indicates extent of enclosure shown in Blackmoor 1569-74, reconstructed both from the first edition OS 6" (1887-9) and fieldwork; other enclosures are not shown. Manors marked thus * are those provided with sufficient names for a boundary survey; other 1569-74 features are as shown..

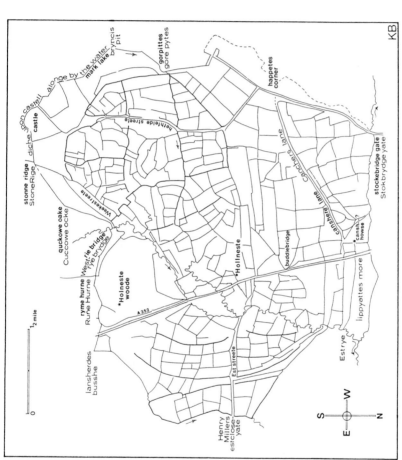

Fig. 2.5 Parish of Holnest drawn from the first edition OS 6" (1889). Names are those given by the Holnest common perambulation (light script) and those by the mapmaker (heavy script). The three marked thus * are in italic, all the others are in a secretary hand, suggesting a coding in the lettering (see note 16). The ring fences, contained within later enclosures, can be compared with those shown for this area 1569-74 (see Figures 2.1, 2.4 and 2.6). Names still in current use include Rhyme Horn, Stonerush Drove (Stone ridge), Galpits Gorse, Stockbridge (Farm) and Cancer Drove.

Common (in Yetminster), and *Queenes common* (Pulham). A single rabbit warren is recorded in Chetnole, *conningres*, probably on the slopes of the knoll of the place-name; and just one hedge, *shiluers hedge*, between Sherborne and Trent. Several gates are named, and these usually divided enclosed land from common. At *stockebridge gate* (Holnest) the mapmaker has drawn the gate, five-barred and coloured orange.

Bridges form the largest category of minor names; over twenty are given. Some take their names from the village in which they were sited, as with *lie* bridge (Leigh), or with reference to the character of their location, *wesworthe* bridge (Folke). *Rie* bridge in Holnest may record an earlier name for the area.[14] Some reflect their structure: *woodebridge* (Holwell); others an association: *pingebridge* (Folke) probably with minnows, *poole* bridge (Caundle Marsh) with an earlier *Deoulepole*, 'devil's pool,' and *hackefordebridge* (North Wootton), probably with hawks.[15] There are personal names in *taylors* bridge (Butterwick), and *bales bridge* (Bishop's Caundle) and *huntingforde* bridge (Folke) seems to record the users, those on horseback presumably taking the ford while those walking could keep their feet dry. *Chitracke* bridge (Caundle Marsh) which carries the present A3030, is of uncertain origin. Thirteen mills are shown, of which nine are in the Yeo valley, but only one appears in association with a named bridge and that is *Oakes* mill beside *West* bridge in Sherborne.

Downstream about a mile and a half from Sherborne there is a river meadow and meander labelled *Charles hammes*. This last name is one of a considerable number which do not appear, at first acquaintance, to belong to any particular category or group. Among these are *cracketayle* (North Wootton), *welles* (Folke), *castle* (Holnest), *foxe hoales* (Sherborne), *vicaries bredde* (Longburton), and *gorpittes* (Holnest), the latter perhaps associated with clay or gravel extraction. There are however, no indications of industrial activity on the map, although medieval pottery and tile-making are evidenced in Holnest, and there were quarries in Longburton, Coombe (Sherborne) and Bishop's Caundle.

Areas particularly rich in names are Sherborne (town) and Castleton, the estates south of Sherborne and thirdly, the area in and around Holnest in Blackmoor which is also characterised by very detailed patterns of settlement and enclosure. Holnest was not only of concern to the mapmaker, but was the subject of a written survey, of broadly contemporary date, entered in the Court Book of the manors of (Long) Burton and Holnest.[16] In short, this entry is one which provides an important clue to the nature and purpose of the cartographic survey of 1569-74, for the written definition of the bounds of Holnest has its complement in the pictorial representation on the map.

In the Court Book the *circute* of the common belonging to the parish of Holnest is described in just twelve lines; the perambulation is completed in thirteen clauses linked by compass directions, and reading clockwise from *Estrye* (East Rye). (In the mid-sixteenth century the greater part of the manor of Holnest was girdled by common; the bounds listed are close to those of the Inclosure Award of 1799/1800, and of the Tithe Award of 1845). Of the thirteen places named, eight are labelled on the map; Figure 2.5 shows the map name in heavy script followed beneath or alongside by the words of the written survey. East Rye does not appear by name on the map, but its location is clearly indicated by a change of colour on the north-east corner of Holnest at the point where the boundary joins the stream. A comparison of the two sets of names provides some interesting complementary information. The word *castle* is found within a dotted sub-circular enclosure which explains the 'bulge' in the present parish boundary at this point. The enclosure itself (since lost) may have been associated with a former coppice compartment.[17] The phrase 'along by the water' of the survey can be identified with the *mark lake* labelled by the mapmaker; and which is thus the boundary of both Holnest and of the Sherborne hundred (to which reference has already been made), and for which the Holnest common perambulation provides an immediate context. Following the written survey northwards we find an unusually long space between *gore pytes* and *Stockbrydge yate*, and the map provides a useful reference mid-way between the two at *happetes corner*, on the edge of old enclosures. The exact course of the boundary here where it runs along Bailey Ridge has been lost following the enclosure of 1800.

Happetes corner is however, the only occasion where the map provides a name not given by the survey, the clauses of which are otherwise fairly evenly distributed around the Holnest perimeter. By contrast, the mapmaker has omitted all labelling along the eastern boundary, although to the observer there can be no doubt whatever that it followed the boldly drawn water course which links East Rye and the corner of Holnest Wood - presumably the site of *lansherdes bush*. Along this length the written survey makes mention of *East Close Gate* providing a mental picture which finds expression on the map. Given that nearly two thirds of the Holnest map names are cited in the perambulation of the common, it seems more than probable that their principal significance lies in association with the Holnest boundary.

The coloured area coincides well with the boundary features shown, and any confusion would arise not along the nameless eastern border, but on the western side where the shading is imprecise. In order to establish as to whether the *castle* lay inside or outside Holnest, it would have been necessary to consult the Court Book. Yet the map has an obvious advantage over the written source in that it provides detail of settlement and enclosure, valuable contemporary information which gives the perambulation a spatial validity - then as now. Careful comparison of the ring-fence enclosures as drawn (Figure 2.6) with their remains still traceable on the first edition six-inch OS (1889) (Figures 2.4 and 2.5) provides ample evidence, not of a fanciful rendering of prettily hedged fields, but a faithful, coded representation - if

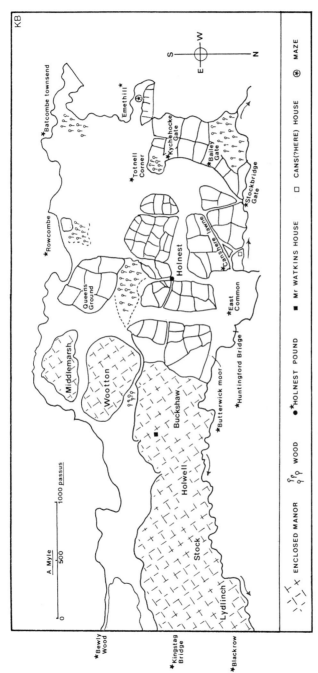

Fig.2.6 Tracing made from the 1569-74 map of the area showing Blackmoor; enclosures can be compared with those reconstructed in Figure 2.4. Names marked thus * are those which define the limits of the lordes praye, some have been modernised (see text). The scale is that given by the mapmaker and transferred from the bottom right hand corner. It may not have referred to the whole map (see note 38).

badly distorted - of agricultural features that actually existed at the time. The question arises as to the precise role of the map. It clearly cannot have stood alone as an estate survey, it need only be compared with maps prepared for neighbouring Hermitage by Norden in 1615 and by More for Minterne and Hartley a year later - eloquent statements of the technical progress made in land measurement and scale drawing, and in the expectations of the landowner by the beginning of the seventeenth century.[19] Such maps were eventually to render the written perambulation obsolete. But in 1570 we may take it that manorial authority rested with the Court Book. Vested in the written word was a legal validity ratified by the practice of many centuries.

The style and content of the Holnest perambulation would not have been unfamiliar to those who drew up a survey of the bounds for the nearby manor of Thornford granted to Bishop Wulfsige of Sherborne nearly six centuries before, about AD 946.[18] The circuit of this estate was completed in just seven clauses linked by compass directions. With several small modifications caused by road and railway construction, these bounds are those of the present parish, and they appear on the 1569-74 map, clearly defined, but not labelled. The relatively few clauses are related to the topographical 'simplicity' of the boundary. The principal east-west lengths are formed the one by a ridge nearly a mile and a half long, and the other by the river Yeo and its tributary consisting of almost three miles of continuous watercourse, and between them receiving only three mentions. As was standard practice, the bounds read clockwise, but there is no indication whatever as to the relative distance between one clause and the next. The number of boundary clauses bears no direct correlation with the length of the boundary, nor with the size of the estate.

In the tenth century the dimensions of the Thornford estate were expressed as a value, not as an area; that is in hides, and not in acres. Six centuries later no value was placed on Holnest, although the surveyor seems to have been constrained to give some indication as to the total distance to be walked, *and by estimacion the circuite of the common belongynge to the parishe of Holnest is ffower myles and all within theyre owne tythinge.* According to the scale given by the mapmaker the total would have been somewhat in excess of five; present measurement makes it very approximately eight. Such divergence is further symptomatic of the fact that the actual length of the boundary was not of over-riding importance any more than the distance between each boundary clause. What mattered was that these clauses should be correctly described in relation to one another, and in the proper sequence, committed to writing according to formula and duly witnessed.

What the mapmaker seems to have provided is a pictorial representation of the parish bound-beating which, it may be noted, had been made a general requirement by royal injunction of 1559.[20] The mapmaker has given graphic substance to a description in words by the labelling of a part symbolic depiction of the landscape to which they were related. Moving across the map - and the scale is large enough to accommodate both the finger and the necessary words - it is a straightforward matter to compose a boundary recital. The compass points are on the borders of the map, and the linking features are all shown whether it be hedge, lane, watercourse or 'notional' - a dotted line across open common. Laid before the informed observer was a powerful *aide-mémoire*, and set before an interested stranger, who had never made the journey, was a picture of Holnest.

No lord's name is given for Holnest, which is not described as a manor. For several centuries Holnest had been an outlying part of the Bishop of Salisbury's Sherborne manor, and this is the most likely explanation for its apparent lack of status. But the manor of Sherborne is itself 'lordless' - an anonymity which suggested to Harvey that the map was commissioned by the Bishop of Salisbury himself. It is of interest to note that in the Bill of Chancery of 1556-58 the Bishop reclaimed rather more than Sherborne Castle. In 1551 Edward Capon, Bishop of Salisbury, had demised to Edward, Duke of Somerset, the manor of Sherborne with the Castle, and the manors of (North) Wootton, Whitfield, (Long) Burton, Holnest, Yetminster, Candel Bishop (Bishop's Caundle), Castleton and Newland (in Sherborne) and the hundreds of Sherborne and Yetminster. Some five years later he declared the lease he had made was 'by threats and for fear of his life'. The Lord Chancellor (then the Archbishop of York) pronounced in Capon's favour and decreed the premises to the Bishop.[21] The estate remained in episcopal hands until 1578 when the Queen obliged the then Bishop, John Piers, to lease it to her. Capon remained in office until 1557, his successor was John Jewel in 1559 (whose escutcheon appears on the south wall of the headmaster's house in Sherborne) and who was followed by Edmund Guest in 1571. Given that the map can be dated to between 1569 and about 1574, it is clear that the lord of the manor of Sherborne was the Bishop of Salisbury, and that at this date Holnest formed part of his estate together with seven other named places and the two hundreds, which, as already noted, occupy the greater part of the map.

Approached in this way, the map can be treated as a portrayal of the episcopal estate. It offers a boundary definition of several of the manors mentioned in Chancery proceedings, not only for Holnest, but for (Long) Burton, (North) Wootton, and Bishop's Caundle. To a degree it refines an understanding of the nature and extent of the Bishop's interests in Whitfield and in the very large manors of Sherborne (including Castleton and Newland) and Yetminster. Thus the field narrows, for the mapmaker has declared his purpose. Paradoxically, the problems of interpretation increase, for it is difficult to know exactly what he was intending to convey, or indeed exactly how much he could have been expected to convey. Complicated patterns of tenure, custom and interest had evolved within the bounds of these two hundreds over many centuries. Such material could not be portrayed on a single sheet, even by the most sophisticated of modern techniques, without the addition of a copious written description. Such a description cannot be reconstructed from the map. But the possibility that such could once have existed provides a sense of direction to future enquiry. What follows is a brief exploration of its potential. In no sense is this a final statement; it is more by way of a progress report on the opening stages of what appears to be, at present, a fruitful approach to the field.

First there is a brief look at the manors of Burton, Wootton and Bishop's Caundle, which seem to have been the subject of similar treatment to that employed for Holnest. Four drawn and labelled 'gentleman's' houses are considered next (although not mentioned in Chancery), followed by a closer look at the manors of Yetminster and Whitfield - the first very extensive in area, and the second today represented by a single farm; the mapmaker suggests the estates were not only adjacent to one another, but were contiguous with Holnest. There follows a very brief consideration of Newland and Castleton, and of the problems presented by Sherborne itself, and we then conclude with that area of country that stretches across the top third of the map, the common of Blackmoor, which, while not specifically mentioned in Chancery proceedings, has nevertheless a link with the manor of Sherborne.

Adjoining Holnest to the north was Burton, latterly Longburton (Figure 2.5). The bounds of Burton may be recited from Stockbridge Gate of the Holnest perambulation, running north along the route through *litle burton* as far as *burton cross*, thence to *vicaries bredde*, a field which formed part of the Burton glebe, where the boundary turns east to follow a small stream as far as *hakefordebridge* to join the bounds of both Folke and North Wootton at what is now Green Lane. Here the Burton boundary turns south to join another stream, and continues on downstream as far as what the Holnest perambulation identifies as Estrye. There is one 'deviation' in the colour (between the two streams) which might have needed a distinguishing label, but otherwise the boundary follows the watercourses which are strongly drawn.

The whole manor of Burton is shown as divided up among a series of fairly regular, uniformly-shaped fields bordered by hedgerows supporting small trees. Neither fields nor hedgerows can be dismissed as 'conventional' even if, as seems to be the case with the fields, they appear to bear little resemblance with those of the nineteenth century. The mapmaker represents hedges in several different ways, and those of Burton need only to be compared with those of neighbouring Wootton and Holnest - the former are shown without trees, whereas those of Holnest are densely packed. While this could describe their actual appearance, it may be significant in other respects; similarly perhaps with field shapes. While empty spaces can sometimes be adjudged under open field cultivation, patterns of enclosure may betray an attempt to represent differing kinds of tenure. In this sense, the enclosed fields shown within the principal Holnest ring-fence, because they broadly relate to those of today, could either be misleading or representational.

As for Holnest and Burton, a boundary recital can be composed for Wootton, which is rendered in a shade of blue-green. Reading clockwise from *vicaries bredde* are a further eight names ranged so as to suggest a direct relation with the manor boundary. First comes *primsley hedd*, and then east to the place where the principal south-bound route from Sherborne reached *gainsborowe hyll* on the Sherborne manor boundary. Beside the road here on the south-west corner of the deer park, the mapmaker has drawn a very small device placed on a hachured line that represents the slope of the hill. This was later the site of *Gallows Plot*, and it seems probable this is a representation of the 'bishop's grim pair of gallows' which, in 1464, lay on the road to *Haukyniscrose*.[22] *Hawkins crosse* is not only labelled on the map, but a small wayside cross has been drawn beside the road that led from the top of Gainsborough Hill down into Wootton. From what is still known as Gainsborough Hill, the Wootton boundary runs east along the deer park pale, clearly shown as a wall (Wootton was 'the village without the park wall' of Walter Raleigh's schedule)[23] as far as *xx acres gate*, thence south following the hedge to *heydon gate*, continuing as far as *cracketayle* on the south-east corner of the manor where the mapmaker has shown a distinctive indentation in the hedgerow. The boundary then turns west to *iiij elme crosse* at the *Anston* (Allweston) crossroads. (The nearby public house is The Three Elms). From here the colouring guides us to *clotfurlongate* to rejoin the Green Way north as far as *hackefordebridge*, where the boundary follows the stream back to *vicaries bredde* beside the present A352.

The map does not say who holds either Burton or Wootton, nor is any name given for Caundle, other than the designation 'Bishop' by which it is still distinguished from the other Caundle-named villages. The Caundle manor is compressed and rather distorted, and there is little room for internal details. The perambulation of the bounds is much assisted by the deep blue-green colouring. This is low-lying land, and the edge of the manor is defined for the most part by bridges, as the bounds zig-zag across country between the small tributary streams of the Caundle Brook. From **kitfordebridge* on the borders of Folke (names marked thus * appear in a recognisable form on the modern Ordnance Survey 1:25,000) there follows **pingebridge*, *bales* bridge, and **poole* bridge, and so east to *wollies cross* on the edge of the village. This latter is probably a family name; the last heiress of the *Le Waleys* married a Fauntleroy of neighbouring Marsh who died in 1440 (see below). Here the bounds change direction yet again, and run north-east as far as *chitrackebridge*. Then they turn east to *idvellgate*, although there is little indication on the map of the pronounced hedged 'funnel' entrance preserved here in the present field pattern, suggesting the location of a one-time gate on the edge of the manor dividing enclosed land from common. The bounds continue in a small hedge that runs along the south side of *holte wood*, on to *garfordegate*, thence south east skirting the slopes of *okill* to join the Caundle Brook at *woodebridge*.

The heavy blue line of the Caundle Brook forms the whole of the southern boundary of the manor, flowing under *corneforde* bridge and powering *corneforde myll*, the only mill shown in this part of the map - probably the Bishop's mill. But unlike the mills of the Yeo valley, it shares with those shown on the River Wriggle south of Yetminster (Chetnole, *stones myll*, and a third unnamed) the distinction of being labelled in secretary hand and not in italic. Upstream, the Caundle bounds continue as far as the confluence at *remslake* bridge, where the bounds leave the watercourse to track across land westwards to include the hamlet of Bishop's Down, and then by way of a place called *welles*, they return north along the road to *kitfordebridge*.

Caplane is the name given to the north-south route that divides Marsh from Folke, but neither estate is listed in the Bill of Chancery, and the mapmaker has not provided enough names for a boundary recital. There is, however, additional information for Folke in the drawing and labelling of *mr Mollins house* (West Hall). This is one of four such houses, *mr Lewstons house* (Leweston Manor), *mr Fauntleroys house* (Fauntleroys Marsh in Folke), and *mr Watkins house* (Holwell) (Figure 2.4). That some selection was involved is suggested by the inclusion of two further 'gentlemen's' houses neither of which is identified by the owner's name. Sir John Horsey inherited the land of Sherborne Abbey manor and his name appears four times on the map, but neither Clifton (Maybank) house, nor the former monastic grange at Wyke bears his name.

A second 'ownerless' house is to be found on the Burton/Holnest manor boundary. The map is badly rubbed here on what was formerly an old fold line; the word *howse* is discernible (in an italic hand unlike the other four *house* names), and the first four or five letters of the first word probably read *cansh*. It stood at the end of *canshere lane*, beside the stream, and at the narrow end of a prominent hedged 'funnel' clearly drawn by the mapmaker, which is likely to represent a one-time stock-herding entrance opening onto the moor from Burton. No road remains on the Burton side of the boundary, and the 'funnel' was, by the time the survey was made, incorporated into enclosed lands in Holnest. On Bishop's land we would not expect to find a personal name. In 1604/5 the greater part of the Bishop's Sherborne estate was granted to Alexander Brett. Included in the terms was 'a liberty of hawking, fishing, and fowling in the manors of Burton and Holnest, late possessions of the Bishop of Sarum'.[24] The house may have been a small hunting lodge; nothing of it remains to view.

The houses of Messrs Mollins, Fauntleroy and Lewston still stand, although much altered. They appear on the map presumably in connection with the Bishop's estate, but as to what the tenurial arrangements were, and they could all have been different, the mapmaker does not say. At the time the map was made certainly these three houses and their lands were 'held of the Bishop of Sarum in the right of his see/as of his manor of Sherborne'.[25] Such holdings highlight the problem as to the meaning of 'manor of Sherborne' in the Bill of Chancery. The Bishop's Sherborne manor was clearly more than part of the large, well defined pale-green territory surrounding the town, and which is labelled as such. But how much else was meant, (and the map may be attempting to say), it is, at least at present, impossible to know. The mapmaker has shown Mr. Fauntleroy's house with a central tower, and Mr. Lewston's house with a chapel attached to it, coloured grey to distinguish it from the residential quarters.

Buckshaw House in Holwell may stand on the site of Mr. Watkins' house. His lands are not well-defined, the house lay on the edge of what was then the enclosed area of the manor, sited close to the hedged 'funnel' entrance to Blackmoor. The common here shares the Holwell colouring, and is named *Queenes Common*. Links between Holwell and the Bishops of Sarum are hitherto slight; the map may constitute useful further evidence.[26]

The great sub-circle of Yetminster land forms a prominent feature on the western side of the map. Not described as a manor, the village of Yetminster with its church is the centre of a large area devoid of both detail and personal names, and rendered in two shades of the same pinkish colouring. There appear Ryme, Chetnole, Leigh and Totnell, but other small settlements are ignored. Yetminster was characterised by a large number of copyholders holding land on long lease; 'the Manner [of Yetminster] is a kind of Fee Simple ... and the Lord hath but a small fine att every death or alienation' noted Walter Raleigh. In 1592 the Queen had conveyed Yetminster to him (as part of the episcopal estate) from Bishop Coldwell 'who was surprised into consent at this alienation and never enjoyed himself after'.[27]

Yetminster was held by the Bishop of Sarum in 1086, and the manor was subsequently divided into four parts, three prebends and the fourth part retained as demesne. The exact extent of the prebends is not established, but the detail on the map for the area around Leigh and Totnell strongly suggests that the Bishop's principal interests in Yetminster were concentrated here. The village of Leigh lay on the north-western edge of the great common of Blackmoor which, at this time, stretched as far as the River Lidden, some six miles to the east. The plots cultivated by the villagers were protected by a continuous back hedge, much of which still remains. A few yards away, at the top of a low hill overlooking the village, the mapmaker has skilfully drawn a representation of a turf maze, by means of a pair of ovals linked by a central cross. It is unlabelled, but there can be no doubt as to its identity. Later known as *Mizmaze*, a low, badly degraded earthwork still remains.[28] The mapmaker has designed a figure to represent a real place which, small as it is, was deemed worthy of inclusion. The question as to whether the chosen figure has any symbolic significance should perhaps be asked.[29]

Land on this side of Yetminster lay adjacent to the episcopal estate at Holnest, and gave the Bishop direct access to, if not control over, two of the principal north-south drove routes of the area; the one running from Stockbridge via Totnell corner, and the other through the *hethfelde streete* ring-fence, and both making the ascent of the chalk scarp above *Stoys gate*, which the mapmaker has duly labelled. A lost route ran direct from Totnell corner south-west towards Holywell on the Ilchester to Dorchester road (the present A37), a short length was incorporated into the common edge near *Batcombe townesende* (Figure 2.6), probably represented today by Sievers Lane.

The route from Stockbridge to Totnell lay along the top of Bailey Ridge (drawn but not named) which divided the Bishop's manors of Holnest and Whitfield. Coloured differently from Totnell, although seemingly sharing the same pattern of fields (or land-use?), Whitfield has a separate mention in the Bill of Chancery. The Bishop's wood at Whitfield is mentioned as early as 1270. The mapmaker has drawn an isolated house here standing within an extensive wood; a sixteenth-century reference to *Whitfeild house* may be this building, of which, at present, nothing further is known.[30]

The Bailey Ridge route is shown running through a wide strip of unenclosed land. For many generations the land here had been held by the Streeche family, who held it 'of the Bishop of Sarum as of his manor of Sherborne by service of five pounds of wax payable yearly at the Feast of the Purification'. Bailey Ridge was the scene of a thrice yearly cattle fair but 'at present [1871] of very little note'.[31] The Bishop's mapmaker has ignored a 'narrow poke' of land here claimed by the Trenchards for their manor of Hilfield in a written survey likely to be of similar date.[32] He has, however, very carefully drawn a narrow pointed 'corridor' of land which gave the manor of Lillington access to Stockbridge (and thus to Bailey Ridge) from the north side. Hutchins notes that both Whitfield and Bailey Ridge were accounted as lying within Lillington. Lillington is, however, not designated 'manor' by the mapmaker, but uniquely as 'market', the *Martket of Lylyngton*. At present the market is perhaps best understood in connection with the Bishop's cattle fair on Bailey Ridge.

The three remaining places listed in the Bill of Chancery are 'the manor of Sherborne with the Castle ... Castleton and Newland'. Lying between the Castle and Sherborne, each of the latter constituted a separate 'Liberty'. The first was a small borough founded in the mid- twelfth century, and the second was laid out by Bishop le Poore in 1227/8.[33] Both owed their existence to Bishops of Salisbury, and both are readily distinguished on the map in terms of both detail and colour. The small manor of *Castletowne* surrounding the Castle and Borough is probably coterminous with that of the old Castleton parish which survived until the end of the nineteenth century. East of the Castle is *St magdalenes* church, (the present church built in 1714 stands on the north-west side), one of only two named churches on the whole map. The other, *St Peters Chappell* , lay near the northern edge of the Sherborne manor in the tithing of Overcombe. Two mills are shown in Castleton, and the borough street, much of which was destroyed by the railway in 1860. East of *Castletowne* are the meadows of *vartnams grounde, blackemershe and the feaver*, the last perhaps the site of the (Bishop's) fish ponds.[34]

Newland is coloured in a deep blue, and includes not only the houses of *Newland streete* but land adjoining to the north-east. There appears to be a pond in the central market area (a *mere* next to the Newland cross is recorded in 1468-88)[35] although the houses, like all the others shown, are portrayed with red roofs, which, in an area of thatch and stone tiles, is unlikely to reflect their actual appearance. Also in deep blue is Honeycomb Wood and *longmeade* on the south side of the river, the hill slopes here show several large irregular enclosures which are traceable today - *guckow hill*, *long hill*, and *honeycombe lease*, which share the pale green of the greater part of the Sherborne manor. The Bishop's deer park is a shade of yellowish green. The Bishop's estate included the park, and by 1603/4 it had clearly included Honeycomb Wood which was made over that year for the use of Walter Raleigh's widow.[36] The blue of both Newland borough and Honeycomb Wood may suggest a link between the two.

The last entry in the Bill of Chancery to be considered here is in fact the first, 'the manor of Sherborne with the Castle'. As already noted, the 'manor of Sherborne' presents very considerable difficulties, and much more work is needed before any kind of interpretation can be made of the mapmaker's treatment of the town and the estate around it. The town itself is shown in some detail, but the manor is characterised by very little detail, and few names. The labelling of *barton barne* on the north-western edge of the town (approximately the site of the later Barton Farm) may perhaps suggest that the Bishop's Sherborne estate was that later known as the 'Manor of Sherborne Barton'.[37]

Without further documentation it would however, be unwise to push the limits of inference too far on any aspect of this map dealing, as it must surely so often, with intangibles in the form of customs, services, rights and privileges. An interesting case in point is provided by another entry in the Holnest and (Long) Burton Court Book and which concerns the greater part of the top (southern) part of the map, which, of its own, gives no hint whatever of any connection with either the Bishop of Salisbury or with the manor of Sherborne.

In addition to the ring-fence enclosures of Holnest, the mapmaker also shows (although in less detail) those of (Glanvilles) Wootton, Middlemarsh, and *Queenes Groundes* in the royal manor of Hermitage. All of these can be traced in the present landscape, although it can be seen by comparing Figures 2.4 and 2.6 that they were misrepresented in terms of relative area, and misplaced with reference to one another. But, as Harvey suggested, it seems probable they depict the state of enclosure at the time the survey was undertaken.[38] With the exception of Holnest however, the mapmaker was clearly less interested in the shape and content of these enclosures than he was with the fact that they were not available for common grazing. These were places, both named and fenced, which had been taken out of the common of Blackmoor.

The bounds of the greater common of Blackmoor are given in a Court Book entry which cannot be securely dated, but which is most likely to date from the middle of the sixteenth century, and thus to be broadly contemporary with both the map and the Holnest perambulation already mentioned.[39] The entry is in the name of John Leweston holding in chief of the Bishop of Salisbury. Within the limits of the common there took place the *praye* or stock-round-up.

> ...all customary tenants of Burtone and Holnest ... ought to
> healpe drive the lordes praye within the compas of the lordes
> royalty and soyle viz From ...

and there then follows a list of fifteen names beginning with *blackrewe* (Blackrow in Lydlinch), *kingstake brydg* (Kingstag Bridge over the River Lidden also in Lydlinch), and *Bewly Wud* (latterly Beaulieu Wood in Buckland Newton) which are arranged more or less north-south along the 'notional' edge of the map - none of them appear (Figure 2.6). From *Bewly Wud* the common edge runs west as far as *rowcombe* above *crocere rewe*; the first is the name of the valley in which Hermitage church stands, and the second is the hamlet known today as Higher Holnest (it was *Weekestreete* to the mapmaker).[40] From here the places are closer together, reading in order, *Batcombe townesende*, and then north and north-east to *Emethill, Totnell corner, kychehocke yate, Bayly gate*, and so on as far as *Stockbridge gate*, thence via *canshere lawne* to *est common, huntingford brydge* and *Butterwyke moure*, finishing up at *Holneste pounde*. A number of these names occur in the Holnest common perambulation already discussed.

This list of *praye* names is readily complemented on the map by a continuous hedgerow which runs from the top left (south-east) corner, along the foot of the chalk scarp until near Batcombe where it turns north. As its course becomes more sinuous the number of names in the list increases; the full complement is shown on Figure 2.6, superposed on a tracing of the map. Of lost names, *Emethill* belongs to a low round-topped hill in the parish of Leigh clearly marked by means of a 'molehill', and the mapmaker has drawn the hedge running across it just below the summit.

In recording the entire length of this hedgerow - and it represents several miles - the mapmaker does not supply any names that seem to apply specifically to it; these names were reserved for the written survey. The map is a geographical *aide-mém-*

oire; the natural complement to its perusal and appreciation was spoken comment. Clearly conveyed is the expanse of open common, and scattered within it islands of enclosed land from which stock was excluded, and around which the *praye* took place. Much of this landscape can be reconstructed. The striking differences in the proportions between one place and another that are the result of transposing the Tudor survey onto an OS map do not, of course, detract in the slightest from the validity of the *praye* in its contemporary pictorial rendering.

> ...and such catail as are within the forenamed praye
> commons and Soyles to drive to Holneste pounde and there
> ...to stay them until X of the clocke the same day and
> none to depart without licence ...

Holnest Pound is shown on the map by means of a small circle.[41] We may take the *lorde* to have been the Bishop of Salisbury.

'Neither one deft blow of a single methodological strategy nor a handy parcel of documents tumbling from a standard archival source is likely to reduce the conundrums of meaning which attach to various types of Tudor maps.'[42] A greater sense of optimism has, however, been provoked by the very positive contribution made to the understanding of this particular Tudor map by two documents from one manor. The potential of future 'parcels' remains to be seen, but the mapmaker has yielded more than enough so far to make it very clear that future enquiry is unlikely to reduce the significance of anything he has chosen to include, or the style in which he has chosen to show it.

Acknowledgements

The author wishes to thank the staff of the British Library, the Dorset Record Office, and Winchester College Muniments for granting access to documents and Kenneth Smith for long-term loan of a copy of BL Add MSS 52522. She is also grateful to Professor Paul Harvey for commenting on a draft of this text. He also very kindly allowed Mrs Barker to read his paper on Tudor estate surveyors in advance of publication. Thanks are also due to J.H.P. Gibb for reading the typescript and, last but not least, to Dennis Seaward for his major contribution.

Notes and References

1. P.D.A. Harvey, 'An Elizabethan Map of the Manors of North Dorset' *British Museum Quarterly*, 29 (1964-5), 82-4, plate xx.
2. The second of the two dates here, 1574, repeats an error made by John Hutchins (*History and Antiquities of the County of Dorset* (1873 iv 501) which states that the Hilfield manor was sold out of Trenchard hands in 1574 when the date actually meant is 1754: see K. Barker and D.R. Seaward 'Boundaries and Landscape in Blackmoor: the Tudor manors of Holnest, Hilfield and Hermitage' *Proceedings of the Dorset Natural History and Archaeological Society*, 112 (1990), 5-22. In the event this makes very little difference; at best the dating of the map could be narrowed by a year to between 1569 and 1573. The latter date can, however only be approximate. If the map were indeed the commission of the Bishop of Salisbury, the only secure terminal date is 1577/8 (see text).
3. P.D.A. Harvey, 'Estate Surveyors and the Spread of the Scale-Map in England 1550-1580', paper presented at the 11th International Conference on the History of Cartography, Ottowa, 1985.
4. See S. Tyacke, 'Introduction', J.B. Harley, 'Meaning and Ambiguity in Tudor Cartography' and P. Eden, 'Three Elizabethan estate surveyors: Peter Kempe, Thomas Clerke, and Thomas Langdon' in S. Tyacke (ed.), *English Map-making 1500-1650* (1983), 13-19, 22-45, 68-78.
5. Bishop Aelfric of Sherborne complained *circa* 1008 to Aethelmaer, Earl of the Western Provinces, that of 300 hides which had contributed to 'ship-scot' in the days of his predecessors, 33 hides had been lost to the Bishopric. In the thirteenth century the Bishop of Salisbury (successor to the Sherborne see) held the hundreds of Sherborne, Yetminster and Beaminster. See F.E. Harmer, *Anglo-Saxon Writs* (Stamford, 2nd edn 1989), 266-70.
6. R.A. Stalley, 'A Twelfth-Century Patron of Architecture: a study of the buildings erected by Roger, Bishop of Salisbury 1102-1139' *Journal of the British Archaeological Association*, 3rd series, 34 (1971), 62-8. Plate xx in Harvey (above note 1) is a reproduction in black and white of the area of the map that shows Sherborne castle, park and town; almost the same area is reproduced in colour in S. Tyacke and J. Huddy, *Christopher Saxton and Tudor Map-making* (1980), plate 77.
7. For a useful summary see M. Aston, *Interpreting the Landscape* (1985), 34-6.
8. E.M.W. Tillyard, *The Elizabethan World Picture* (1943), 91-108.
9. K.V. Thomas, *Religion and the Decline of Magic* (1971), 117-18

10. A.G. Dickens, *The English Reformation* (revised edn, 1967), 225. Regarding the purchase of church lands he notes that 'The compunction, the sentiment, the nice scruples, the superstition that an evil destiny awaited the buyers, these were invented by later and more poetic generations'.

11. A sixth escutcheon was probably that of Bartholomew Combe of Lillington who left money in perpetuity for the remuneration of the school usher; see *The Sherborne Register* 1550-1937 (Cambridge, 3rd edn, 1937), xxvii.

12. Part of this map is reproduced in D. Burnett, *Dorset Before the Camera* (Wimborne, 1982), 120. It has been suggested that each hundred of Armada Dorset (1588) was responsible for setting its own beacon; only 8 hundreds have no recorded beacon site (of which Sherborne is one). *The Beacon on bubdowne hill* (labelled by the mapmaker in italic hand) lay in Yetminster Hundred. See F. Kitchen, *Fire over Dorset: the Armada Beacons* (Brighton, 1988), 6, 10 and 14. Hill-side fires continued to be lit during and after the Reformation on St John the Baptist (24 June) or St Peter's eve (28 June); see Thomas (above note 9), 82. The mapmaker has marked 'St Peter's Chapel' (now lost) beside one of the two principal routes north out of Sherborne manor, close to its highest point.

13. A.D. Mills, *The Place-Names of Dorset* (English Place-Name Society vol. 59-60, 1989) Part 3, 334.

14. *Ibid.*, 338

15 *Ibid.*, 332, 317 and 382.

16. See Barker and Seaward (above note 2) for full text transcription, discussion and references for this Holnest survey. Both the entry in the Court Book and the names given for Holnest by the mapmaker, are in versions of the secretary hand, the style used for all routine recording and correspondence. The exceptions for Holnest are *Holneste woode, Hollneste* and ?*cansh ... howse* which are written in the italic or chancery hand, recommended by Mercator for its clarity (Figure 2.5). See A.S. Osley, *Scribes and Sources: handbook of the chancery hand of the sixteenth century* (London and Boston, 1980), 20 and 187. Both styles are distributed over the whole map, and presumably relate to the perceived 'status' of the various names. While spelling is perhaps more related to the sound of the name, the variations given for Holnest may be of some other significance.

17. The Holnest 'castle' is one of three such enclosures in this area. See Appendix 2 of Barker and Seaward (above note 2).

18. 'A plott ... of certain groundes ... of the Princes manor of Fordington' John Norden, 1615 (Dorset Record Office Photocopy 362 (fo. 13)), and 'The Topographical description of Minterne and Hartley ... bi John More ... 1616' (Winchester College Muniments no. 21378).

19. G.B. Grundy, 'The Saxon Charters of Dorset: Thornford' *Proceedings of the Dorset Natural History and Archaeological Society*, 60 (1938), 87-9; P.H. Sawyer, *Anglo-Saxon Charters* (1968), no. 516.
20. M. Beresford, *History on the Ground* (1957), 28-9.
21. Hutchins (above note 2), 214
22. J. Fowler, *Mediaeval Sherborne* (Dorchester, 1951), 124 and 147. This was probably the site of an execution in 1261: also Mills (above note 13), 215.
23. Hutchins (above note 2), 215
24. *Ibid.*, 217
25. *Ibid.*, 128, 179 and 181
26. *Ibid.*, 521; also RCHM, *Dorset Central* (1970) vol 3, part 1, 118 and 121.
27. Hutchins (above note 2), 214 and 215.
28. K. Barker, 'The Mizmaze at Leigh, near Sherborne Dorset' *Proceedings of the Dorset Natural History and Archaeological Society*, 111 (1989), 130-2.
29. See remarks by Harley (above note 4), 34-5, who also notes that if a maze is to be shown on a map it can only be meaningfully represented by a plan. This particular symbol at Leigh is capable of being interpreted as the two hundreds, the one oval within the other and bound together by the Cross; or the figure '100' read twice, once horizontally, and once vertically. The Roman numerals 'CC' can also be extracted.
30. Mills (above note 13), 344.
31. Hutchins (above note 2), 196 and 197.
32. 'Mem- of a View of the Range of Hilfield Common', see Appendix 1 in Barker and Seaward (above note 2).
33. Fowler (above note 22) ,146-64.
34. Mills (above note 13), 311.
35. *Ibid.*, 364.
36. Hutchins (above note 2), 215.
37. *Ibid.*, 220.
38. Harvey (above note 1), 83. The scale-bar provided by the mapmaker has been placed on Figure 2.6. A comparison of Figures 2.5 and 2.6 brings out the distinct 'narrowing' of Blackmoor as the 1569-74 map approaches the River Lidden and the edge of the survey. To draw, to a fixed scale, a map of a single estate confined within a rectangular border, is to undertake (if such has not already been done) a considerable amount of measuring of other people's land - the first step towards the making of an 'atlas'. The scale-bar is not a formal affair, and only lightly drawn in, but it presumably appears for a reason. It might eventually prove possible to demonstrate that it only applies to selected areas of the map.

39. Dorset Record Office D721A/1; for a full reference see Appendix 1 of Barker and Seaward (above note 2). The *lordes praye* is found after entries of 1551/2, before the Sherborne Custumal of 1582, which is followed by entries of 1552 and 1553.
40. For the identification of *rowcombe* and *crocere rewe* see Appendix 3 in Barker and Seaward (above note 2).
41. The first edition of the Ordnance Survey six-inch series (1888) shows Holnest Pound in a very similar location to that suggested by the Tudor mapmaker. The latter has also drawn a pound at the southern end of Burton village.
42. J.B. Harley (above note 4), 23.

Fig.3.1   John Norden, the Townships of the Soke of Kirton, Lindsey, Lincolnshire, the only surviving portion of a larger map.   Reproduced by gracious permission of His Royal Highness, Prince Charles, Duke of Cornwall.

CHAPTER THREE

Patronising the Plotters: the Advent of Systematic Estate Mapping

Graham Haslam

This paper seeks to trace the history and use of estate plans by the Duchy of Cornwall from the last quarter of the eighteenth century. After 1770 estate maps rapidly became first an important and then a crucial tool for management of a large, diverse estate. A landed estate such as the Duchy of Cornwall controlled a variety of properties, foreshore, fundus, minerals, farms, urban land and property undergoing transformation as it was developed. Estate maps offered an effective tool which could provide immediate and accurate information for management of real property assets. The emergence of this technique, which became by the end of the nineteenth century usual in a majority of legal instruments dealing with real property transactions, fundamentally affected the ways in which landlords could manage their property. By the twentieth century the estate plan became essential and definitive for all real estate transactions. The important place of the estate map today suggests two obvious questions. When did maps become a crucial part of management technique and is the Duchy typical now and in the past of the way other landowners viewed property management?

It is easier to take the last question first. The Duchy has always been a large landed estate. Today it extends to about 128,000 acres located in a dozen counties. In the past it has owned even more acres in as many as seventeen counties. In Dorset the Duchy has owned Fordington, which surrounds the market town of Dorchester on three sides, since its foundation in 1337. Additionally, it possessed the manors of Ryme Intrinsica and Ryme Extrinsica and in the later nineteenth century purchased Gillingham. Other major holdings were in Surrey, Lincolnshire, Somerset, Devon as well as in Cornwall. Since its foundation the Duchy has always owned land scattered across the face of England south of the Trent.

The Duke of Cornwall is the heir apparent. When a duke does not exist then the estates are vested automatically in the crown and administered from the Exchequer, becoming an independent estate only when there is a Duke of Cornwall old enough to govern. However, from the reign of Charles I onwards Princes only indirectly managed the Duchy's affairs. From the end of the fourteenth century until the nineteenth century the Duchy had perpetually absentee landlords. No Duke of Cornwall visited Duchy lands from the time of the Black Prince until the nineteenth century but that does not mean that management could not be rigorous and detailed. In the periods when a Duke existed the Duchy was governed by the Prince's Council, a wholly London-bound body which was dependent upon a loose network of crown servants to supply it with what we would call 'management information'.   In the occasionally very long periods when no Duke existed the Duchy was given over to the Land Revenue side of the Exchequer, assigned to one of the seven auditors of the Land Revenue and fell more or less wholly into the orbit of the Lord High Treasurer.

Was the Duchy typical of landed estates? A somewhat circumspect answer is that the Duchy was typical of its kind. In fact, it was representative of one very important category of landed estate. As is well known, most property was in the hands of a very small percentage of the population. Estimates vary because little in the way of reliable data exist before the middle of the nineteenth century, but it is clear that estate owners possessed a very significant share of the landed wealth of this country.

In order to make sense of estate ownership it is necessary to categorise them in such a way as to explain variations in economic, social and political aspirations and management. We will ignore the *rentier, t*he small owner of a farm or copyhold who partially or wholly rented his land to one or more individuals in order to secure an investment income.   Here only the estate must be defined.   That word became prevalent in the reign of Elizabeth I to define the real property holdings of certain individuals.[1] The clear implication of the usage of the word 'estate' is of rights of ownership in land without direct involvement in cultivation of it.

Virtually all estates may be described as belonging to one of three categories. They may be conceived according to the possible ways and means they could be managed by their owners; the categories are not necessarily gradations of wealth. In other words, it is possible for somebody to have an estate in the first which is actually worth more than an estate in the second category. However, normally it would be expected that these categories reflect gradations of wealth as well as social and political status. Naturally, some estate owners were recipients of wealth which did not originate from their acres. The Gore-Langton family built up their estate at Newton Park just west of Bath in the seventeenth and eighteenth centuries while they remained prominent Bristol merchants.[2] Men moved relatively slowly from trade to landed gentry.

To define the first and most modest, but most numerous, category: it consisted of estates where owners resided more or less permanently. On an estate of this kind, it would be expected that owners rented much of their land, and occupied and worked a farm, perhaps with the name of 'manor' or 'home' farm. The estate would be contiguous or in close proximity with all the land, rented or directly farmed, located as a single parcel. In the 1770s the Cornish nabob, John Call, arrived home from India laden with the fortune he had amassed while with the East India Company. He naturally aspired to the life of a country gentleman, purchased property in the Tamar Valley, not far from Stoke Climsland, and built a fine country seat, Whiteford House. Because the amount of land which he could purchase freehold was limited, he also bought many copyhold estates belonging to the Duchy which were adjacent to his property in order to create a larger contiguous estate of his own.[3] Eventually, when the Call estate suffered the same fate as the Gore-Langton, the Duchy was able to re-acquire the copyholds and buy the freehold land and the house. On estates such as these the owner would hardly need what we would now call formal management information. He could, and did, go and look for himself whenever problems arose. He knew his tenants personally and was involved in their social as well as economic affairs. If you like, we can think of Squire Weston or perhaps Squire Allworthy as archetypal, if fictional, examples of this class or group of estate owners. They could be efficient or inefficient with only a modicum of formal data because they were resident on the land which they owned, lived amongst their tenants and to some degree shared their aspirations and their problems. It is overwhelmingly an image of comfortable wealth, of status without aggrandisement. These landed gentlemen might spend a season in London, but their interests were tied firmly to the affairs of their estates.

The next category includes intermediate estates. Within this group owners possessed properties which were not contiguous; their lands might be in different counties. Each property might consist of a different manor so that the customs and property law for each could well vary to a greater or lesser extent. However, the owner of an estate of this kind could and usually did visit all his properties regularly. An example would be an eighteenth-century family such as the Kenyons, who came to wealth in part through the practice of the law. They owned lands in three counties and regularly spent summers on their Lancashire property, moving south to their Flintshire lands before the rigours of the more northern winter descended.[4] On each estate they possessed a substantial residence where they spent a part of each year. They may be said to have direct knowledge of the affairs of all their tenants and their lands but owners of this kind of estate could not be in two places at the same time; it required something like a more formal management structure and more detailed records. The properties themselves were topographically and geographically varied and needed different sorts of management decisions. It is quite likely that a steward or other officer would have assisted the estate owner in the provision of management information and ensured that the privileges of the owner were safeguarded. Some kind of formal record structure would have been necessary to ensure that rents, tenures, common rights and other complex matters were properly noted. In social terms the owners of these estates might well be knights of the shire or even, as in the case of the Kenyon family, ennobled. They were often important county figures, but their interests gave them greater scope and wider vision than those estate owners whose lands were in a single location.

The final category consists of those estates which were very large and scattered over a wide area so that the owner found it impossible to visit all of his lands annually. He was thus by circumstance at least a partially absentee landlord; in an extreme case, such as the Duchy of Cornwall, he might even be continuously absentee. These estates often had a considerable and complex historical pedigree, dating from the great medieval baronies, modified by a succession of marriage alliances which amalgamated and dissolved the original estates to re-cast them in a series of changing, but enduring matrices. Though these very large estates were fewest in number, they nevertheless shared a vast amount of property and controlled in the pre-industrial era a very considerable share of what we would call the gross domestic product. The enormously wealthy upper strata of English society often became leading patrons of the arts, architecture and literature. Upon this kind of estate some form of management structure was essential if the owner was to maintain control of his prerogatives and privileges.

Defined in this way, a specific number of acres is not crucial to the characterisation of each category in the model. The Duchy of Cornwall was typical of the largest, scattered estate. It did possess some characteristics which other large estates did not share; the fact that it was for administrative reasons merged with crown lands when no Duke existed is one very important difference between it and other estates of comparable situation. However, like them, it was essentially an entailed estate which provided income for its incumbent, whether Duke or monarch, from rents and fines. Because no Duke of Cornwall could sell capital assets, which were its lands, it actually provides a more refined, easier case study than other estates where the entail could be broken and capital realised from sales. Capital movement considerably clouds a study of both management technique and financial performance of an estate.

Another important distinction between the Duchy of Cornwall and other very large estates is that no great house became associated with it. Eighteenth- and nineteenth-century Princes enjoyed building, of course, but each Prince created his own folly without encumbering future Dukes of Cornwall by erecting on the estate a vast country seat which would generate continuing costs of maintenance as well as need numerous servants to make it comfortable. This distinction was probably not important until the beginning of the twentieth century when large houses became a catastrophic burden upon estate owners.

It is necessary to consider the particular history of the Duchy of Cornwall in order to establish how and for what purposes its lands were administered in the seventeenth and eighteenth centuries so that it is possible to determine how valuable maps were to its management. This seemingly long road to the map collection of the Duchy will explain why estate plans were created.

In the Jacobean period the Duchy underwent considerable administrative change but it was a quick and relatively painful transition. The details are beyond the scope of this paper, but once the patterns of administration were set, they remained unaltered until another wave of reform broke over the Duchy in 1840. Hence for almost two and a half centuries methods of work remained fundamentally unaltered. From the early seventeenth century the Duchy was governed by the Prince's Council which was appointed directly by the Duke. Its size could vary, but it essentially consisted of courtiers, duchy officers and legal men. Duchy officers were usually, but not invariably, the chancellor, the lord warden of the stannaries, who was also chief steward of all Duchy manors, the receiver-general, who acted as rent collector and banker, the attorney-general, who provided legal advice, the solicitor-general, who processed leases and copyholds to which the Duchy was party, the Prince's secretary, who obviously marshalled correspondence received and acted as clerk to the Prince's Council to prepare outgoing correspondence and other written material, and the surveyor-general.

The surveyor-general was in fact a Jacobean revival of an office which had not existed for over a hundred years and its functions are central to our story of estate maps. The office of surveyor-general dates back to the late middle ages and a Court of General-Surveyors existed in the early sixteenth century to administer the crown estate.[5] It was eventually amalgamated with other Exchequer departments, but the crown estate (more correctly, the Land Revenue side of the Exchequer until finally reformed in the eighteenth century) continued to appoint surveyors for each county or sometimes groups of counties. This the Duchy did not do until Prince Henry Frederick, eldest son of James I, revived the office and appointed a courtier, Sir William Fleetwood, to do the job.[6]

Any tenant, anybody who wished to become a tenant, take a lease, dispute a decision of a local Duchy officer, or transact almost any other kind of business, did so by petitioning the Prince's Council which was granted full financial and administrative powers of the Prince's lands. Prince Henry, and later Prince Charles, when his elder brother died, commissioned surveys of those lands. Fleetwood did not personally carry these out but rather retained, on a piece work basis, a number of deputies who actually surveyed the properties. John Thorpe was one, but the most active deputy by far was John Norden.[7]

These deputies carried out surveys of particular manors in the ancient form. Armed with a commissioning warrant, they journeyed to the manor and empanelled an extraordinary manorial jury. The deputy surveyor then put a series of questions to it, such as what are the bounds of this manor, did mills exist within the manor, and such other questions as satisfied the needs of the lord of the manor. A surveyor could also call for all relevant court rolls so that he could compare the answers of this jury with its historical precedents.

Norden was, as distinct from those who worked with him, a pioneering cartographer. His famous county maps represent a considerable technical achievement in the history of cartography. Besides an evident fluency in the specialised forms of Latin used in recording land law, he also used surveying instruments such as the theodolite and chain when he worked for the Duchy.[8] He was, however, working in the time-honoured way. When he concluded the examination of the extraordinary manorial jury, and walked the manor, he then produced a report which is known generically as a terrier. The Oxford English Dictionary defines 'terrier' as a register of landed property, formerly including lists of vassals. This in essence was exactly what Norden was doing but he provided more information in these terriers than his predecessors would usually have done. For instance, he detailed real land values as opposed to nominal, he provided a lengthy, often acerbic commentary upon each manor and a detailed discussion of tenants' obligations and privileges, real and imagined, and other factors which influenced the economy of the locale.

Norden understood that careful surveys of property could lead to benefits for a landlord. This he argued at length in his work, *Surveyor's Dialogue.*[9] He had the technical capacity to produce scale plans of lands, and of a whole estate if requested. He was clear about the benefits which such technical surveys could bestow both for providing a clear legal definition of the owner's title and in bald economic terms.

If we now revert to the model of English landowning constructed earlier, it can clearly be seen that Norden's services would be very attractive to those owners who could not visit and personally supervise the affairs of their properties. The last two categories of estate owners delineated in the model would be most interested in the new technology offered by Norden. An estate such as the Duchy of Cornwall, managed by a Council tied to London, could reap obvious benefits from his work. Clearly, the Duchy was interested in his services; he and his son, also John Norden, worked as deputies for over ten years compiling a detailed series of terriers which eventually included virtually all manors except for Dartmoor.[10] Norden produced maps for the royal family and most well known of these is the plan of Windsor. This is an exquisitely produced item on vellum, a virtuoso piece, in part an advertisement for his work.

The Nordens' terriers were far more detailed and provided much more information than we would expect of a medieval terrier. All the acreage measurements were clearly derived from the use of a chain and theodolite and they are presented in a tabular format. As such they would have provided the basis, the processed data, for maps and in at least one case Norden did produce a plan to accompany his terrier. This is a map of the Soke of Kirton in Lindsey in Lincolnshire (Figure 3.1), a property of over 21,000 acres.[11] The whole plan is not extant, nor is it signed but it is clearly by Norden. The cartouche is characteristic of his work, as is the colouring; this is a typical example of Norden and furthermore very much in the mode of his Duchy work. The plan is oriented east/west, not north/south and depicts a series of townships with their associated lands. At the time the plan was produced, the Prince's Council was considering the sale of the soke and were interested in total acreage, current values per acre and conditions of tenure. Norden pointed out in his accompanying terrier and observations that the capital value of the land would be greatly enhanced if the fens were drained and he does in fact indicate the position of a drainage dike on the map.

Obviously, river communications were important and both the Humber and the Trent are prominently depicted. The lord of the Soke had specific fishing rights and other kinds of ownership over parts of the Trent. The road network is clearly delineated, the position of towns and hamlets, particular houses, and churches are all included. There is one windmill marked on the map which also indicates fens by use of dot symbols, pasture by green colour and let demesne land is edged red. On the whole this plan is a magnificent production, certainly worthy of that eccentric master. The extant map represents only a fraction of the original which was severely damaged at some indeterminate time in the past. The soke extended almost to the gates of Lincoln so a very great deal is now missing.

This map is the only surviving Norden plan of Duchy properties. It is possible that other manorial maps were made by Norden for the perusal of the Council, but if that is the case all have been lost. Contemporary references to Norden's work by Duchy officers are to his 'surveys' and do not specifically mention plans, maps or charts, but continue to see the survey in its generic, medieval context.[12] The wonder of the technical revolution, as distinct from its economic benefits, was lost on the Jacobean Prince's Council.

Norden's work, however, quickly became the abiding reference for all property investigations. It continued to serve the Duchy until the late eighteenth century. In the Interregnum, parliament undertook surveys of Duchy property but these are much more abbreviated than Norden, sometimes they do not even contain acreages. Most of these surveys were carried out between 1649 and 1652 by Exchequer officials.[13] The intent was to sell the lands of 'Charles Stuart' in order to raise money quickly to pay an army growing increasingly riotous. Thus there was no need to provide detailed information, but only to convey to potential purchasers what was on offer. They are akin to what today we would call 'estate agents' particulars'.

When the Duchy was re-established by Charles II, again surveys were undertaken, but these were even more terse than those ordered by parliament and drew upon Norden for much information.[14] Their prime intent was simply to re-establish the crown's authority to title. Without significant exception the Duchy continued to appoint a surveyor-general through the late seventeenth and for virtually the whole of the eighteenth centuries. Each petition to the Prince's Council for a copyhold, renewal or reversion or for a leasehold was referred to the surveyor-general for a report. He constructed a title history of the property, provided information about past rents and fines and made a recommendation to the Council on the offer received from the petitioner. There is an unbroken series of his reports from 1660 to the end of the eighteenth century.[15] Nevertheless, with a few honourable exceptions, almost none of these refers to a plan or makes use of an existing chart. Norden had provided both technological expertise and a systematic approach which in practical terms proved lucrative. Yet this approach was all but abandoned for what was essentially a medieval method. The reason for this is not just to be found in the vagaries of social and political events, but lies imbedded primarily in the nature of property law and what that meant for tenants and landlords alike. Most Duchy land was let by copy of the court roll according to the custom of the manor and by the late seventeenth century this was stereotyped to three lives. Sometimes tenants opted for a ninety-nine year lease, but this was actually for three lives or ninety-nine years. Leaseholds were usually for thirty-one years.

This meant first, that for long periods property was out of the hands of estate owners who had no responsibilities relative to the land for the time property was vested in leasehold or copyhold. Without a financial stake in buildings, equipment or livestock in agricultural areas, it effectively meant they could not intervene in the affairs of their tenants with planned regularity. It is true that renewals, reversions and petitions to add lives were frequent, but they were received on an irregular basis related to the immediate needs of the petitioner rather than according to any pattern which might conform to a landlord's strategy for his estate. In urban areas the situation was quite different, but before 1750 the Duchy did not possess any significant urban land.

There was little change before the eighteenth century in the legal relationship between tenant and landlord. However, the agricultural revolution provided land-lords with an impetus to seek improvements to their lands so that they could then demand higher fines from their tenants. In 1761 a Wiltshire gentleman, Edward Baynton of Spye Park, was appointed surveyor-general.[16] Eventually given a baronetcy, he continued in office until 1795. He worked closely with a deputy, Richard Gray, who actually supervised the day to day work of the surveyor's office.[17] There was by this time urgent pressure on the Duchy to provide as much income as possible as Prince George, eventually George IV, was by nature a profligate spender.

Additionally, his relationship with his father led to a series of arguments which caused an embarrassing and more or less irreconcilable breach between them. The Prince lost all financial support other than the Duchy, only because it was something George III could not deny his son.

Baynton and Gray retained a west country surveyor, William Simpson, who began to survey Duchy lands in 1771. Though a fairly obscure figure, he seems to have come from Chippenham and been an established estate surveyor before he began working for the Duchy. He may have been known to Baynton as it is possible he worked at Spye Park. He continued to work for the Duchy until his death at the end of the century and in all he produced twenty-eight estate maps.[18] All of these, in colour and on vellum, were produced to a very high standard. Like his predecessors, he was commissioned to survey a manor, visited it and empanelled an extraordinary manorial jury to which he put questions. Like Norden's work, which he examined closely, the product of his labours was much more sophisticated than the traditional survey.

To accompany every plan he produced a terrier.[19] The two were linked by assigning numbers to each enclosure so that it was possible to locate an area on the map quickly and efficiently. Simpson's terriers provided details of acreage, the place name and use of each and every enclosure within a manor and, additionally, he provided information concerning the names of all tenants. Perhaps most importantly, he provided either a foreword or afterword in which he summarised the condition of the manor, crops and what improvements might be rendered, including enclosure, to bring about more income. Typically, these detailed terriers run to eighty or ninety pages for each survey. Simpson continued in Duchy employ for over twenty years and his last work was in 1799 when he died.

The plans have virtually all the elements of a modern plan. The manor, washed with colour, appears to float as an island on the tan background of parchment because the maps stop at the manorial bounds. The maps are enormously detailed and contain furrow lines, pasture, orchards, natural features such as rivers and meres and boundaries, such as hedgerows, walls and fences. After his death, Simpson's brief was taken by Thomas Davis who added accurate depictions of buildings in the first few years of the nineteenth century as he carried out re-surveys of the properties.[20] There is no doubt that these maps, created for the purposes of mundane estate management, brought the countryside into the heart of London and the very chamber where the Prince's Council met.

It is especially interesting that Simpson did not include buildings upon his plans. This is a cautionary tale; the researcher using manuscript estate maps must always consider the reasons why the map was created. Landlords had no interest in the eighteenth century in fixed capital assets and so buildings were irrelevant to their purposes. They could persuade or be persuaded to change field boundaries and land uses, but the shape of the farmstead was of no consequence.

Simpson's career as a deputy surveyor, spanning thirty years, gradually included more and more work. In the decade of the 1770s he produced five plans, all estate maps; in the next decade he surveyed and mapped six Duchy properties (Figure 3.2) and one estate which was not part of the Duchy. In 1788 he produced a plan of Sutton Pool, the harbour for Plymouth, with the clear intention of providing a definition of title boundaries (Figure 3.3).[21] Another map, that concerned with the Duchy's estate of Berkhamstead in Hertfordshire, included only the castle, park and demesne lands within the manor.[22] Clearly, his skills were being used for more specific purposes. In his final decade of work, the 1790's, he produced fourteen plans, but these included only eleven separate estates. For Mere, Wiltshire, he produced two separate manorial plans, one completed in 1798 and another in the next year.[23] For the manor of Shepton Mallet he plotted both a comprehensive manorial map and a detailed plan of the houses and gardens within the manor.[24] In 1797 he mapped Prince's Wood and Bromby Warren located within Kirton in Lindsey (Figure 3.4).[25] Simpson worked mapping Duchy estates and properties in Wiltshire, Somerset, Dorset, Hertfordshire, Lincolnshire and Devon.

In Cornwall, where the Duchy owned almost 30,000 acres of land, Baynton and Gray engaged another surveyor, Henry Spry. He was possibly a son or brother of John Spry of Boyton and may have been related to William Spry who produced a manuscript map of Canada and the St Lawrence in 1760.[26] Spry knew Cornwall well. The earliest map he produced for the Duchy is dated 1784 and from then until 1790 he produced sixteen manuscript maps of thirteen estates (Figure 3.5).[27] From the very beginning his work was sometimes specialist. In 1784 he provided a plan of the Harewood Estate within the manor of Calstock, the demesne at Bradbridge and the castles and parks of Launceston and Restormel.[28] In the decade of the 1790s he produced a further seventeen plans concerned with sixteen estates. Interestingly, Spry delineated an index imposed upon a map of the whole county of Cornwall.[29] His final manuscript map is dated 1797. In addition, there exist three maps without date which could belong to either decade.

Like Simpson, Spry used surveying instruments and produced terriers to accompany his plans but his terriers are not as detailed as those of Simpson. He listed tenants and produced accurate acreages, but he did not provide synthetic comments concerning the condition of the land. However, his work did contain an important technical departure. Adjoining the descriptions of farms in each manor, he placed a miniature of his larger plans so that a user had both an accurate plan and the terrier side by side in what amounted to an atlas for the estate.[30] The estate plan had at last achieved equal status with the ancient written terrier and become an integral part of it.

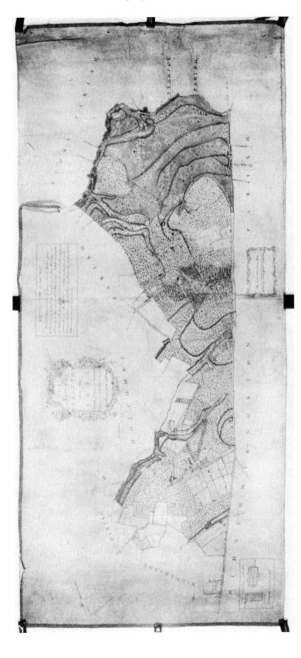

Fig.3.2 William Simpson, the Manor of Stratton-on-the-Fosse, Somerset. Reproduced by gracious permission of His Royal Highness, Prince Charles, Duke of Cornwall.

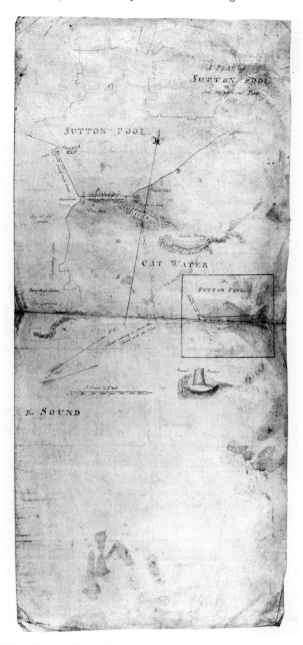

Fig.3.3   William Simpson, Plan of Sutton Pool, part of Plymouth harbour, Devonshire. Reproduced by gracious permission of His Royal Highness, Prince Charles, Duke of Cornwall.

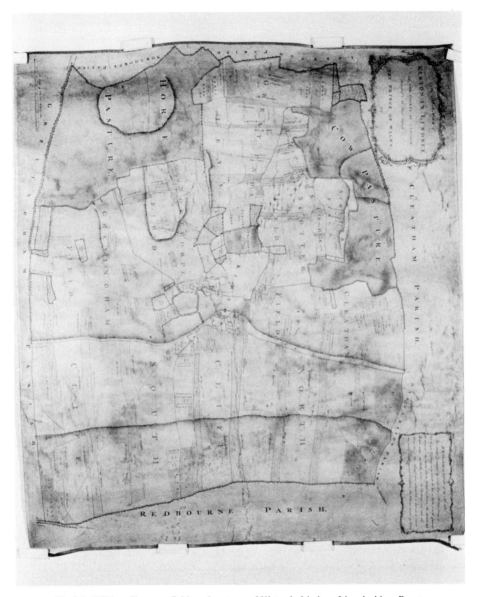

Fig.3.4 William Simpson, fields and pastures of Kirton in Lindsey, Lincolnshire. Reproduced by gracious permission of His Royal Highness, Prince Charles, Duke of Cornwall.

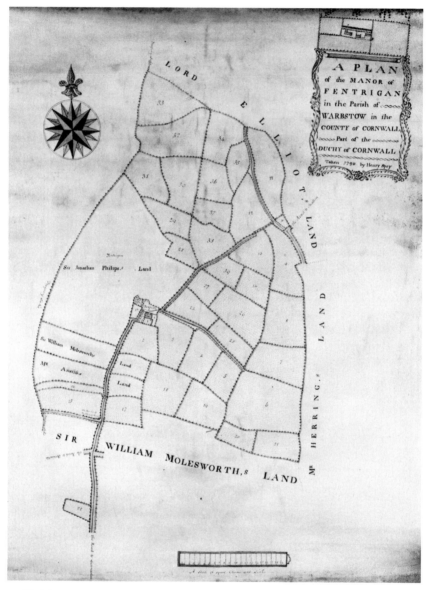

Fig.3.5   The manor of Fentrigan in the parish of Warbstow, Cornwall, surveyed in 1790, is one of sixteen manuscript maps made for the Duchy by Henry Spry and is fairly typical of his work. Though in the same style as Simpson, it lacks the fine colouring and buildings were not added. Reproduced by gracious permission of His Royal Highness, Prince Charles, Duke of Cornwall.

In the remarkable period between 1770 and 1820 there existed two other estate surveyors who deserve consideration. The Duchy owned the manor of Kennington on the south side of the Thames up-stream from the City of London. By the last quarter of the eighteenth century this manor had two well-known features. The first was Vauxhall Gardens, created by Jonathan Tyer, a kind of eighteenth-century Disneyland where the fashionable came to see and be seen amidst a rococo fantasy. Kennington was also an intricate network of market gardens providing fresh vegetables for London's tables. Its traditional copyholds had been divided and sub-divided into a complex of precisely measured and jealously guarded plots. The Oval cricket ground, located within the manor, was one such market garden area in the eighteenth century.

This whole area was transformed from market garden to suburb as the great bridges were thrown over the Thames. These came one after the other from the middle of the eighteenth century until the Waterloo Bridge Company, incorporated by parliamentary act in the first decade of the nineteenth century, actually provided direct access to Kennington. Urbanisation and the busy wharfs built along the foreshore of the River provided the Duchy of Cornwall with significant income and opportunity for more. In 1785 two surveyors, Messrs Hodskinson and Middleton, were contracted by Baynton to survey the manor.[31] They laboured for over twelve months to produce a veritable street atlas of Kennington which is coloured to show demesne, leasehold and copyhold lands. Additionally, every building is depicted at a scale much larger than on J. Rocque's celebrated map of London.

From the end of the eighteenth century, the Duchy retained one surveyor after another to continue the production or modification of estate maps. As landlords such as the Duchy of Cornwall became involved directly in financing agricultural fixed capital assets, surveyors' services became virtually indispensable. In 1829 the Duchy hired George Driver, founder of Drivers Jonas, whose father had started surveying when, as a market gardener in Kennington, he found a better living measuring market gardens rather than growing crops. George Driver was retained to survey the most westerly of all Duchy property, the Isles of Scilly.[32] The terrier is in five volumes but the plans, though extant, are no longer a part of the Duchy's archive.

In 1844 the Duchy of Cornwall hired its first full-time surveyor as a direct employee.[33] From then until now the Duchy has been divided geographically into districts and a surveyor, known as a land steward, placed in charge of each. However, from the 1840s there was a rapid decline in the production of manuscript estate plans. The great day of the estate surveyor as map maker had passed. Very quickly Duchy employees came to rely on tracings from first the tithe maps, and after 1880, the six-inch scale Ordnance Survey maps. Estate plans simply became copied extracts from comprehensive national plans. These were and are accurate and tidy, but lack the flair and individual genius which a long line of Duchy cartographers had produced for almost three centuries.

Notes and References

1. J.A.H. Murray (ed.), *The Oxford English Dictionary*, (1897), vol. 3, 300, definition 11.
2. Duchy of Cornwall Office (hereafter D.C.O.), Title Deeds, Newton Park Estate, boxes 1 to 14.
3. D.C.O., Title Deeds, Whiteford Estate, O1 and O3.
4. L. Stephen and S. Lee (eds.), *Dictionary of National Biography* (1885-90), vol 11, 30-2.
5. W.C. Richardson, *History of the Court of Augmentations, 1536-1554* (1961), 141-4.
6. D.C.O., Rolls Series, Prince Henry's Patent Rolls, Box 28, nos. 539 and 540.
7. *Ibid.*
8. John Norden, *The Surveyor's Dialogue,* (1607), 56.
9. *Ibid.*
10. D.C.O., Bound MSS, Letters and Warrants, 1626-1632, fo. 28.
11. D.C.O., Maps and Plans, unnumbered plan of the Soke of Kirton in Lindsey, R/B/1.
12. D.C.O., Bound MSS, Acts of the Council,1622-1623, fo. 6.
13. N.J.G. Pounds (ed.), *The Parliamentary Survey of the Duchy of Cornwall*, Devon & Cornwall Record Society, New Series vols. 25 and 27.
14. For example, see, D.C.O., Bound MSS, Surveys, 1660-1661, by Price and Prideaux, S/2.
15. D.C.O., Bound MSS, Reports of the Surveyor-General, 1660-1813, P/M/2 and 3.
16. D.C.O., Bound MSS, Patents and Warrants, 1714-1760, G/B/2, fo. 160.
17. D.C.O., Calendar, Eighteenth-Century Administrative Papers.
18. D.C.O., Maps and Plans, nos. 977 (Curry Mallet, Somerset), 1071 (Stratton-on-the-Fosse, Somerset), 664 (Stoke under Hamdon, Somerset), 1190 (Fordington, Dorset), 1261 (Kirton in Lindsey, Lincolnshire), 1260 (Berkhamstead, Hertfordshire), 1953 (Sutton Pool, Devon), 1133 (Longbredy, Dorset), 586 (Langton Herring, Dorset), 1263 (Yaddleton), 1030 (Shepton Mallet, Somerset), 576 (Inglescombe, Somerset), 617 (Widcombe, Somerset), 666 (West Harptree, Somerset), 1032 (Laverton, Somerset), 1134 (Farrington Gurney, Somerset), 1910 (Garthorpe, Leicestershire), 1205 (Mere, Wiltshire),

1043 (Bradninch, Devon). In addition, Simpson re-surveyed a number of
these manors in order to deal with specific problems.
19. For example, D.C.O., Bound MSS, Surveys, S/46, M/M/3.
20. This is contained in D.C.O., Bound MSS, Surveys, S/44, M/M/3.
21. D.C.O., Maps and Plans, nos. 741.
22. D.C.O., Maps and Plans, no. 1260.
23. D.C.O., Maps and Plans, nos. 1205 and 875.
24. D.C.O., Maps and Plans, nos. 1030 and 1192.
25. D.C.O., Maps and Plans, no. 1261.
26. R.V. Tooley, *Tooley's Dictionary of Mapmakers* (1979), 582 and also, G.C.
Boase, *Collectanea Cornubiensia* (1890), 918.
27. D.C.O., Maps and Plans, nos. 73, 1876, 1189 and 1925.
28. D.C.O., Maps and Plans, nos. 979, 1876 and 1925.
29. D.C.O., Maps and Plans, no. 653.
30. For example, D.C.O., Bound MSS, Surveys, Surveys of Cornish Estates, S/8,
M/M/1.
31. R. Edwards, 'Some Notes on the 1785-1786 Survey of the Manor of Kennington by Joseph Hodskinson and John Middleton' in A.L. Saunders (ed.), *London Topographical Record*, 25, 131-8.
32. D.C.O., Bound MSS, Surveys, The Isles of Scilly, 92-96, R/M/3.
33. D.C.O., Parliamentary Accounts, 1845.

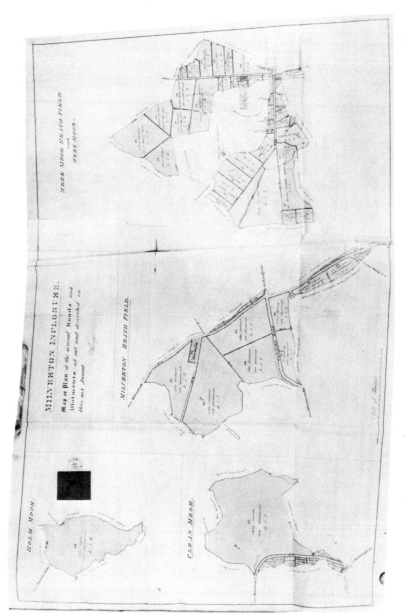

Fig.4.1   Milverton 1831-33. An enclosure covering several widely scattered areas of land, with no indication from the map of the relative locations of each area. However, this map does indicate the owners of adjoining old enclosures, allowing the new holding structure to be investigated in more detail. Reproduced by permission of the Somerset Archive and Record Service: reference Q/RDe 99 (Milverton).

CHAPTER FOUR

The Interpretation of Enclosure Maps and Awards

John Chapman

The interpretation of enclosure maps and awards presents considerably greater problems than, for example, that of tithe maps, reflecting the much more complex and long-drawn-out process which they recorded. The enclosure process took place over many centuries, and even the more restricted parliamentary phase lasted for three hundred years, during which period ideas, not only on how the process should be carried out, but also how it should be recorded, underwent substantial changes. Enclosure maps were not regarded as an essential part of the process during the earlier phase, and when they did begin to be produced there was no standard pattern which had to be followed. The local surveyors and commissioners were left free to indulge their own particular whims, and a great range of styles evolved, some of them more visually attractive than useful as a source of information. Some stuck strictly to the portrayal of the areas being enclosed; others depicted the whole parish, including houses and gardens and private lands of no relevance to the enclosure. Ownership of land was sometimes shown on the map and sometimes not. As a generalisation, one may say that the maps must be taken together with their accompanying awards in order to abstract information of real value.

Legally, the enclosure movement took two forms. If all those with an interest in a piece of open or commonable land consented, the land could be enclosed by agreement, either in accordance with some scheme drawn up by the parties themselves, or by employing outsiders, variously described as valuers, commissioners, or arbitrators, to carry out the redistribution on their behalf. If, on the other hand, agreement could not be reached, the process could be initiated by act of parliament on a petition from some of the interested parties, normally representing the owners of at least two-thirds of the land. In this case, the enclosure was carried out by a variable number of outside commissioners, sometimes supported by surveyors, quality men, and arbitrators. The commissioners themselves often had some say in this, for they usually named the surveyor, and the Exton, Somerset, act permitted them to appoint quality men if they wished.[1] Precise details tended to be a matter of local custom, often county-based. Somerset, for example, regularly appointed arbitrators to hear appeals against the commissioners' verdicts, whereas the other south-western counties did not, preferring appeals to the Quarter Sessions instead. There was also a variation through time, with early enclosures employing large numbers of commissioners, twenty at Wick Rissington and thirteen at Prestbury for example, while later ones favoured three until 1801, after which a single commissioner became the norm.

In general terms, enclosure by agreement preceded parliamentary enclosure, though the two overlapped chronologically, with the parliamentary phase beginning in the seventeenth century and private agreements straggling on well into the nineteenth. Nationally, Wordie has argued that enclosure saw a peak in the seventeenth century, almost exclusively by private agreement, though there was undoubtedly a significant earlier phase in the fifteenth century and enclosure did not cease in the sixteenth.[2] In the South West there is ample evidence of an active enclosure movement prior to the eighteenth century, and in some areas much of such open field as existed was long gone before 1700, while major inroads had been made into the common waste.[3] However, the South West also saw a significant number of agreement enclosures in the eighteenth century and south-western examples occur amongst the more limited nineteenth-century ones.[4]

In spite of the substantial overlap, there was a general tendency for enclosure under parliamentary authority to replace enclosure by agreement from about 1720 onwards, though some early acts may be misleading. Up to and including the 1720s acts were often used to give full legal authority to enclosures already carried out by private agreement, on occasions long before.[5] However, the number of such acts is small as a percentage of the total, and they tend to be largely restricted to a handful of Midland counties and to the early years of the movement, though there is a need to be wary of stray examples elsewhere. The Didmarton and Oldbury, Gloucestershire, award of 1830, for example, refers to an agreement enclosure of 1789 covering part of the same area, though in this case there appears to have been a complete reorganisation, since it was felt that the original attempt had not been wholly successful.[6]

As far as the parliamentary enclosure movement is concerned, the South West saw what is generally regarded as the earliest (Radipole, Dorset, under an act of 1604) and one of the latest (Chipping Sodbury, Gloucestershire, finally completed in 1908), so it was involved in the whole time-span. However, if attention is focused on the main period of large-scale use in the region as a whole, the South West would have to be characterised as a relatively late adopter of parliamentary enclosure. The parliamentary enclosure movement shows evidence of a diffusion process from the Midlands, and the peripheral nature of the South West ensured a time-lag.[7] On the other hand, there were considerable variations within the region, for much of Gloucestershire and Wiltshire were close to the centres of diffusion and were affected relatively early, while at the other extreme Cornwall was not involved before the nineteenth century.[8]

The basic distinction between parliamentary enclosures and those by private agreement applies not merely to the actual process, but also to the record which they left behind. Though some agreement enclosures mimicked very closely the form of the parliamentary awards, and indeed were sometimes carried out by officials experienced in the parliamentary process, such cases are relatively few in number and almost invariably eighteenth or nineteenth-century in date.[9] Prior to 1700 enclosure maps are very rare indeed, and even the written record is usually brief and uninformative compared with a parliamentary one. Some may occur as a mere passing reference in the records of the manorial court, while others may even be disguised as something entirely different, as, for example, in the case of Fratton, Hampshire, whose enclosure is concealed as an agreement to rectify manorial boundaries.[10] Many commons and open field systems simply disappeared from the record, to reappear at some later date as enclosed fields held in severalty.

The records of the parliamentary phase of the movement are in general far better and rather more uniform than those of the private phase, though even here there are substantial variations, partly brought about by changes in legislation, but also reflecting in part local and individual habits and quirks. Until 1836 each such enclosure had an individual act of parliament, but legislation in that year made an individual act unnecessary if certain specified conditions were met. In theory this legislation applied only to open arable land, and was not extended to include common and waste until 1840, but it appears to have been misapplied in some cases. In the context of the South West, this opportunity was seized upon particularly in Dorset, where sixteen enclosures made use of it, and these, of course, have left no act as evidence of their existence.

Later legislation of 1845 reinstated the need for direct parliamentary sanction, though enclosures under this were merely listed in batches in consolidated acts. These were presented to parliament by a newly-appointed Inclosure Commission, which was given the responsibility of authorising the enclosures and ensuring that they were properly carried out in accordance with parliament's guide-lines. This procedure, also, was much used in the South West, though in this case most strikingly in Cornwall, where such enclosures represent some 58 per cent of the county's total.

Even for parliamentary enclosures, the official record may vary a great deal, and may be remarkably scanty. No specific document is necessarily present in every single example. As has already been indicated, no act exists for enclosures such as Bere Regis or Caundle Marsh, both in Dorset, since authority for them derived from the General Act of 1836. Others such as Upper Swell and Tetbury, both Gloucestershire, have no award, the first because the act merely authorised the Lady of the Manor to enclose land which was presumably already all in her own hands, the latter because it vested the former commons in trustees rather than dividing them amongst right owners. More importantly, early enclosures were only rarely accompanied by maps. Since the lands obviously had to be surveyed before enclosure could take place, it would seem logical to the late twentieth-century mind that a map should have been produced. However, accurate map-making was a slow and expensive process, and skilled cartographers were not always readily available in the early part of the eighteenth century. As a consequence, early commissioners almost invariably made use of a written survey of the existing layout as a basis for the re-distribution, and not infrequently simply described the new allotments in terms of their 'metes and bounds' as well, rather than producing a map. Nineteenth-century enclosures are usually accompanied by a map of the new situation, but pre-enclosure maps remained relatively uncommon, unless it was possible to make use of one which already existed for another purpose.

It follows from this that there are enormous variations also in the amount and type of information which can be abstracted from each enclosure. However, there are certain broad groups of information which one might expect to be able to get from most.

A starting point is the land or lands being enclosed. The act obviously needed to specify to which area the powers to enclose applied, and equally obviously the award had to record precisely which piece of land had been subjected to which decision. Unfortunately, the position is not always so clear as this might imply. The act, in particular, had to be worded in such a way as to allow for all possible eventualities, and those drafting it were concerned to ensure that there could be no legal challenge on the grounds that a particular piece of land fell outside the terms specified. Thus while some acts refer very precisely to, for example, 'a certain common called ...', others merely allude vaguely to 'certain open and commonable lands lying in the parish of ...'. Still others listed all possible types of land which might be enclosed, regardless of whether or not they existed within the specific parish concerned. If they did not, no harm would be done, since land which did not exist could not be enclosed; on the other hand, problems would be avoided if, for example, land used as common arable were to be adjudged subsequently to be legally part of the common waste. Confusion may also have arisen from the habit of using earlier acts as a model. Landowners or agents unfamiliar with the enclosure process merely copied successful examples, substituting their own local details where they judged necessary. The Avebury, Wiltshire, act was certainly used in this way as a model for West

Tanfield, Yorkshire, for correspondence between the Earl of Ailesbury, a prime mover in both, and his agent in Yorkshire still exists.[11]

It was a common custom for acts to give an acreage for the land to be enclosed, but these figures must be treated with great caution. The majority are inaccurate to a greater or lesser degree, and the level of inaccuracy can often be extremely high.[12] In a 10 per cent sample of West Country awards, only five of the act estimates were within an acre of the figures produced by totalling the individual allotments, against thirty-eight which were further away. Fourteen differed by more than 100 acres, and the Martock, Somerset, act underestimated by almost 580 acres, or well over a third. Additionally, thirty-three did not even give an estimate.[13] In these circumstances it is not possible to get any real picture of the amount of land enclosed in any area from the acts.

Perhaps more surprisingly, the figures given in the awards are also highly suspect. Since the land had, by then, been surveyed, it might seem reasonable to expect them to be accurate, but this is by no means always the case. Some merely repeated the act figures, for example at Bitton (the 1819 act) and at Longhope, while those which provided a different one might still be wildly astray. Only three of the south-western sample proved to be within an acre, as against forty-nine which were not, and fourteen (not necessarily the same enclosures as in the case of the acts) differed by over 100 acres. The Exton, Somerset, award, the extreme case, differed by 880 acres. Again, the position is made worse by the problem of missing figures, with twenty-three of the sample awards offering no estimate.[14]

A further logical assumption, that the award figures should be more accurate than those in the acts, is again not borne out in practice. Of the twenty-seven sample enclosures where both act and award provided a figure, and these differed, fourteen act estimates were closer to the real total against only thirteen award ones. While nationally it seems that the award estimates are more likely to be accurate, it is certainly very unwise to assume that this is so for any particular individual case. In the final analysis the only way of obtaining accurate figures for enclosures is to total the areas for each individual allotment from the award or, if they are shown, from the enclosure map, an extremely long and laborious task for many larger awards.

Even when this is done, problems remain. Quite apart from occasional discrepancies between areas on the map and in the award, or between the figures given in the text and in the marginal summaries in some awards, there is the problem of what the total represents. Enclosures were not merely about enclosing land which was open or common at the time; in many, reallocation or exchange of lands which were already held 'in severalty' (not subject to any common rights or restrictions) played a significant part. At Deerhurst, Gloucestershire, for example, almost 780 acres was redistributed in this way, out of just over 2500 acres involved. It is therefore important to distinguish between the over-all total of land affected by an enclosure, a total of land allotted by the commissioners, excluding private exchanges, and a total for land which had formerly been open or common, excluding both private exchanges and also old enclosed land given up for the commissioners to re-allocate. Some of the discrepancies already mentioned between act and award estimates and totals by addition arise from this: in the Biddlestone, Wiltshire, enclosure, for example, the act figure (293 acres) appears to be an estimate of the total of open land (244), while the award figure (721 acres) appears to refer to the total involved (731).

It may be noted in passing that not all the land mentioned in an act or award was necessarily enclosed. Particularly in later enclosures, commissioners had powers to leave areas common, either for recreational purposes for local inhabitants or because the land was judged to be of too poor quality to be worth the expense of fencing. Owners themselves sometimes requested that land should be allotted in common, usually where the amounts due to each were very small. South-western examples of land left common at enclosure include Turnworth, Dorset, where a small area was left common to a group of owners, and St Columb, Cornwall, where just over 408 acres was similarly left as a common pasture.[15]

The parliamentary enclosure movement was principally concerned with reorganising patterns of ownership, and the determination of who received what is one of the more straightforward aspects of the interpretation of enclosure documents. There are occasional minor problems as, for example, when the allottee is described simply as 'The Owner', presumably because the commissioners were unsure to whom the title legally belonged, but this is rare. Potentially more confusing is the not uncommon situation where the original claimant died before the award was completed. It was then the normal practice to allot the land to 'the Representatives of ...', or sometimes to the trustees by name. The heir or heirs, meanwhile, might well be allotted land either in their own right or by purchase. Since the commissioners were usually aware of the true situation, the layout of the holdings might reflect this knowledge, giving what, at first sight, might appear an unnecessarily complex and illogical distribution. It would also give an impression of a larger number of smaller holdings than was in fact the case. In view of the long time-span which sometimes passed between the passing of the act and the completion of the award, such occurrences are not infrequent. Some discrepancies between map and award also appear to arise in this way, though more usually this is due to late sales or exchanges after the map had already been drawn up.

Ideally one would like to be able to compare the before and after situation in order to determine how different owners were treated and to discover how far the efficiency of the holding layout had been improved. The former aspect has been a matter of some dispute, for it has been suggested that some owners were treated with less consideration than others, either in the location of their new holdings or even in the quality and amount that they received.[16] The latter has received little attention, for it has tended to be assumed that a greater efficiency of layout would automatically follow the enclosure process. In practice this was not necessarily so. Commissioners rarely succeeded in compacting holdings into a single block, and it was quite normal to scatter even fairly small allotments in several discrete parcels, often some distance apart.[17] This was presumably done in the interests of fairness, but it did involve the production of a pattern of landownership which was less efficient than could have been achieved in theory. However, there was little that the commissioners could have done to overcome one problem, namely that many enclosures involved several scattered commons or fields, rather than a single compact block. This occurred, for example at Milverton, Somerset, where the commons to be enclosed were widely scattered about the parish (Figure 4.1). In cases such as this, many owners might have entitlements to parts of different commons, and the legal position might be doubtful unless the act had been drafted with great care. It was not always clear whether the commissioners had the authority to compensate an owner for his rights in several commons by an allotment in one, and many commissioners preferred to play safe and split the allotment, thus producing a fragmented post-enclosure pattern.

The pattern of ownership was complicated by two factors, firstly land sales, and secondly exchanges. Sales of allotments which were arranged privately between individuals are normally recorded in a straightforward manner. If an owner sold his land or rights between the submission of a formal claim and the allocation of land in response to it, the commissioner normally recorded the land under the name of the new owner, but with a note that his right to it was derived 'by purchase from ...'. Additionally, however, some awards made land available to would-be buyers from another source, namely 'sale allotments'. Faced with the problem of how to cover the costs of the enclosure process, the majority adopted the idea of a rate levied on each individual recipient of land in proportion to its value, but others sold some of the land being enclosed, normally at open auction. This was a particularly common feature where the land concerned was open common and waste, rather than open field, so it was frequently used in the South West. In the sample referred to earlier, more than half the Devon awards and more than three-quarters of the Somerset ones made use of this method of raising costs, and in the case of Somerset this resulted in over 10 per cent of all the land enclosed being put up for sale in this way. Treatment of such sale allotments varies, but it was normal to list them together either at the beginning of the main part of the award or, less usually, at the end, with the price paid for each also normally recorded. Caution is needed, though, for a few awards bury them in the body of the text, where they can be overlooked, and there is evidence, though not as yet from the South West, that small sales of low value were simply lost in an owner's other allotments.[18]

Where this type of information exists, it can be of great value. It is obviously possible to determine those individuals who were actively buying-up land in an area and those who were selling, and to determine whether the buyers were those who had already had a stake in the land or were outsiders. It is also possible to throw light on land prices, and their spatial and temporal variations.

Exchanges of land may cause confusion, not least because they are not recorded in a consistent manner, and because the awards occasionally present them in highly convoluted wording (Figure 4.2). As has already been indicated, some of the land involved in this process was already enclosed; indeed, one of the commonest procedures was for the recipient of a new allotment to exchange it for an old enclosure situated elsewhere, and not necessarily even in the same parish. Old enclosures, in fact, became involved in an enclosure in two separate ways. On the one hand, some old enclosed land was simply thrown into the general pool, to be re-allocated as the commissioner saw fit, though with the owners obviously receiving appropriate compensation.[19] Some awards stipulated that any detached portion of old enclosed land of less than a specified size, usually three acres, or any land owned by two separate individuals but lying within a single ring-fence should be re-allotted. In this case the land was dealt with in the same way as the common, and indeed it is not always possible to decide how much land was of which type, particularly when a single allotment includes both a piece of common and an old enclosure. Some awards, however, record them precisely, and even include the name of the former owner, as at Awre, Gloucestershire, which specifies 'heretofore the property of...'.[20] Such arrangements are occasionally presented as exchanges, though they were so only in the sense that the owner would receive some other land in compensation.

The second type of involvement of old enclosed land was that of exchanges in a more genuine sense, where two owners privately agreed on an exchange for their mutual benefit, and requested the commissioner to include it in the award as an official legal record of the transaction. How it was recorded varied. The normal procedure was to list such exchanges separately at the end, repeating the details of any ordinary allotment involved in the exchange. Other awards included them in the body of the text, recording details under the name of the new owner, while others recorded the whole transaction twice, once under each party to the exchange, for example Locking, Somerset.[21] In these circumstances the possibilities for confusion are considerable, and accidental double counting of the areas is a real danger.

Old enclosures in this sense must be distinguished from encroachments. The latter had been filched from the common at some time previous to the enclosure, and were technically illegal. Treatment of these lands varied. It was usual to specify a time limit, most frequently twenty-one years, but up to forty years on odd occasions: any encroachment older than this was held to be the property of the encroacher, anything

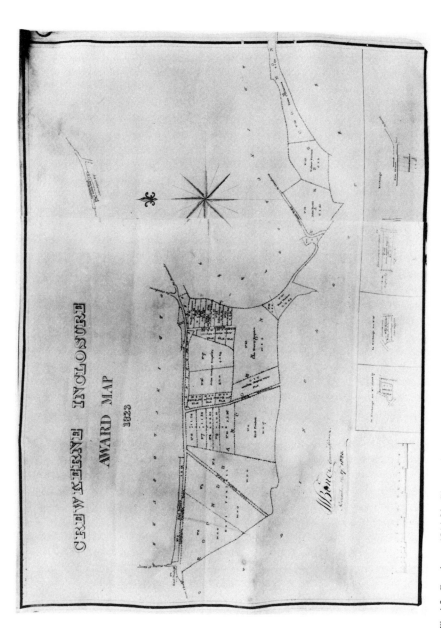

Fig.4.2 Crewkerne 1814-23. Though the original commons are clearly shown, as well as the new allotments, the locations of the old enclosures exchanged are difficult to determine. Reproduced by kind permission of the Somerset Archive and Record Service: reference Q/RDe 26 (Crewkerne).

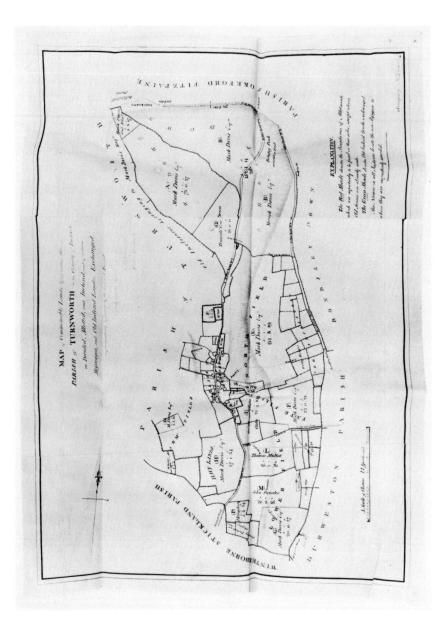

Fig.4.3 Turnworth 1801-5. This map shows very clearly each individual field and common being enclosed, as well as the new allotments and their owners. Unlike Milverton, the owners of adjoining old enclosures are not normally given. Reproduced by permission of Dorset County Record Office: reference Encl. 1.34.

younger being part of the common. At Abbotswood, Gloucestershire, for example, twenty-one encroachments were disallowed, though very many more were regarded as legally established. For the former the owners received no compensation.[22] Treatment of disallowed encroachments was again far from uniform. Some enclosures, such as Sturminster Newton, simply allotted them in the same way as any other part of the common, the encroachers having no special rights to them.[23] Others, such as the Bitton enclosure of 1827, sold the encroachments to raise costs, and in this case it was usual to offer the encroacher the chance to buy privately.[24] Only if he declined was the land offered for public sale.

The possibility of determining the types of land involved, and of reconstructing the former boundaries of individual fields or commons varies greatly from enclosure to enclosure. Clearly, detailed and precise reconstruction is only possible where maps exist, but the presence of a map is not necessarily a guarantee that anything of value can be produced. Some maps, such as that for Abbots Wood, show nothing more than a series of allotment boundaries and key numbers, so that unless the accompanying award is both detailed and helpful there is little that can be done without the aid of other sources such as tithe or estate maps.[25] Fortunately, many awards do in fact specify in some detail the former name or names of the land contained in each allotment, though where a single allotment straddles the boundary of, for example, a common field and a common pasture, the proportions lying in each are very rarely indicated. Other maps are far more helpful, and show not only the boundaries of the allotments, but also those of each of the former fields, commons and meadows which were being reallocated (Figure 4.3). For some of the earlier enclosures, this is the only source which we have of precisely where the fields once lay, and totalling the areas for the allotments described as forming part of a particular former field or common is by far the most reliable evidence of the size of each unit.

Information on the occupation and residence of those involved is tantalising: there is sufficient to suggest interesting possibilities for investigation, but not enough to allow a detailed and comprehensive analysis. It would be of great interest to know who the allottees were: how many were locals, and how many were absentee landowners; how many belonged to the squirearchy and gentry; how many were yeoman farmers; and how many were tradesmen or craftsmen with no apparent direct involvement in the land. For some awards this type of information is provided in full, for example Long Ashton, Somerset.[26] For the majority, however, no such details are given, while still others, infuriatingly, give details for some owners but not others. Fortunately, a reasonable number of awards give details of one of the more interesting and significant groups, those who bought the sale allotments, so at least it is often possible to throw some light on one of the other controversial aspects of enclosure, namely who were the relative gainers and losers from the process.[27]

Care must be taken when generalising from information about occupation and residence, for those awards which contain this information are not a random sample. Nationally, at least, those most likely to contain full details seem to be those from the earliest and latest periods of enclosure. A high proportion of awards under the General Act of 1845 contain residence, often indeed the full address, and many also record occupation or status. At the other end of the time scale, those before the Napoleonic War period are more likely to contain these details than those during it, though there are many exceptions on both sides. It is probable that these comments apply equally to the South West, suggesting that Cornwall ought to be better covered than Somerset, for example.

One subject on which the enclosure awards remain regrettably reticent is the identity of the officials responsible for the process. The commissioners are always named, for they were legally responsible for the proceedings, and all powers were derived from them, but surprisingly in many cases the name of the surveyor or surveyors is omitted, even from the map. The commissioners usually appear in the original act, and any changes which took place between its passing and the completion of the award are carefully noted in the text of the award. Thus William Osborne of Thornbury died during the Horton, Gloucestershire, proceedings and was replaced by John Biedermann of Tetbury, a relative of one of the other commissioners, the German-born Henry Augustus Biedermann.[28] The identification of the commissioners is of some significance, for though most of them acted for only a single enclosure, during the peak of the process a small group of individuals became effectively full-time professionals, and dominated it in particular regions. In the South West these professionals were represented by men like Richard Richardson, operating at various times from Devizes, Bath and Lincoln's Inn, who undertook at least fifty-two of his seventy-seven awards within the region, and Francis Webb, of Stow and Salisbury amongst many addresses, who acted as commissioner in at least thirty-three enclosures within these counties. Care must be taken over the identification, for several commissioners founded dynasties which continued for several generations, and identical names may recur. Richardson followed his father, a Darlington-based commissioner of the same name who somehow became involved in one Gloucestershire award, and several of the Webb family were involved in enclosure work. Another leading West-Country commissioner, Thomas Browne of Sevenhampton and various other addresses in Gloucestershire, shared his name with at least nine other commissioners, though in this case none appeared to be related to him and most operated elsewhere in the country.

Individual commissioners approached the task in their own way, determining how the awards were presented, and probably having their own favourite ways of laying out the holdings.[29] The full extent of their influence as landscape planners has never been assessed, though there are studies of their work in certain areas of the country.[30]

The status or occupation of the commissioners is normally noted, as well as their residence, and this can throw interesting light on the origins of the profession. Tanner, schoolteacher, and iron-master are amongst the occupations recorded, though the majority were drawn from the local gentry and yeomanry. Unfortunately, later awards became stereotyped. The mere fact of being a commissioner conveyed a certain status, and it became normal to refer to them as 'gent.', even though they were not necessarily gentry in the accepted contemporary meaning of the word.[31] Later still, well into the nineteenth century, professional competence became a matter of significance, and a new convention of referring to them as 'land surveyor' began to be adopted. While the conventions say something of the changing perception of the commissioners, it means that only in the earlier awards is this type of material worthy of serious analysis.

The post of surveyor was even more one for the professional, but the absence of information from so many awards makes it difficult to appreciate the full importance of particular individuals. In some cases, one of the commissioners doubled as surveyor, but it was frequent practice to appoint a separate individual, usually a nominee of the commissioners. For many, work as a surveyor was a kind of apprenticeship before they moved on to become commissioners, but others, such as Edward Webb, brother of Francis, specialised in surveying throughout their careers.

Mention of the other major officials by name is unusual in the awards, and they must be sought in the more ephemeral documents, often with little chance of success. Most powerful was the clerk to the enclosure, who was capable of wielding great influence, especially if the commissioners were weak or inexperienced. The clerk was frequently a local solicitor with good connections amongst the major land-owners of the county, in which case he might be employed time and time again on different enclosures. Christopher Pemberton of Cambridge, whose private papers happen to have survived, developed a considerable enclosure business in Cambridgeshire, and there must have been similar less-readily-identifiable figures elsewhere.

In conclusion, some comment is necessary on the present whereabouts of the enclosure documents. Surprisingly in view of their legal status, until 1845 there was no standard arrangement for their preservation or inspection. The most common arrangement specified in the acts prior to that date was that one copy of the award, and the map if any, should be enrolled with the Clerk of the Peace of the relevant county, and it was sometimes specified that the original award should be deposited in the parish church. The Lord of the Manor was sometimes to receive a copy, and some early awards were enrolled in the Court of Chancery. Those enrolled with the Clerks, and usually the parish copies, have now found their way to the respective County Record Offices, while the Chancery copies are now in the Public Record Office in Chancery Lane. Unfortunately, the instructions in the acts were not always followed. The clerks to the enclosures were frequently very lax in following the instructions, and a few awards never found their way to the specified place.

The various associated papers involved in the enclosure movement, the commissioners' minute books, the accounts, the statements of claims, and the various more ephemeral materials are strictly speaking beyond the terms of reference of this account. Nevertheless, it may be noted in passing that their survival has been at best erratic, and their present location difficult to find. It was not felt necessary that such material should be preserved once the process had been formally completed, and documents have survived largely by chance, because no-one felt sufficiently moved to throw them away. Most have been discovered amongst the private papers of individual commissioners, of land agents to one of the major parties involved, or of the firms of solicitors who provided the clerks and various legal services to the enclosures. Some of what is left has now found its way to the County Record Offices, but it is likely that others still survive in private hands.

After 1845 the position was more straightforward. Copies of awards and maps for all enclosures authorised by the Inclosure Commissioners were to be deposited with them.[32] Via the Board, later Ministry, of Agriculture, who later took over the powers of the Commission, these records have found their way to the Kew branch of the Public Record Office. Greater care was also taken with much of the supporting documentation, and this too is now usually to be found at Kew. Unfortunately, the great majority of parliamentary enclosures had already taken place before these arrangements were implemented, though this late final phase was of considerable significance to some parts of the country, notably Cornwall.

In summary, there are a great many potential pitfalls in interpreting the enclosure maps and awards, and study of them discloses many irritating gaps in the record. Nevertheless, they present us with an enormous mass of data, much of it not yet properly analysed, about one of the most significant upheavals in the history of the English landscape. Concentration by so many workers on the English Midlands has left parts of the South West little studied from this point of view, and a thorough investigation of the enclosure documents would greatly enhance our knowledge both of the development of the area and of the nature of the enclosure process as a whole.

Notes and References

1. Quality men, as the name implies, determined the relative values or qualities of different parts of the land to be enclosed so that this could be taken into account in the allotment. This work was normally done by the commissioners themselves.
2. J.R. Wordie, 'The Chronology of English Enclosure, 1500-1914', *Economic History Review*, 2nd Series, 36 (1983), 483-505; M.W. Beresford and J.G. Hurst, *Deserted Medieval Villages* (Lutterworth, 1971); J. Chapman, 'Changing Agriculture and the Moorland Edge in the North York Moors, 1750 to 1960' (unpublished M.A. thesis, University of London, 1961), 33-64.
3. E.C.K. Gonner, *Common Land and Inclosure* (1912); W.G. Hoskins *Devon* (2nd Edition, Newton Abbot, 1972); H.S.A. Fox, 'The Chronology of Enclosure and Economic Development in Medieval Devon', *Economic History Review*, 2nd Series, 28 (1975), 181-202.
4. See R.E. Sandell, *Abstracts of Wiltshire Inclosure Awards and Agreements* (Wiltshire Record Society, 25, 1969). A late example is Wanstrow, Somerset, award 1842.
5. For example, Farmington, alias Thormarton, Gloucestershire, 1713.
6. Wiltshire R.O., Q/RI 54.
7. J. Chapman, 'The Extent and Nature of Parliamentary Enclosure', *Agricultural History Review*, 35 (1987), 25-35.
8. Gloucestershire's first was Thormarton, 1713; Cornwall had none before 1809, and only 3 out of 33 before 1836.
9. For example, Great Wishford, Wiltshire, 1809, used John Seagrim (4 parliamentary awards) and Christopher Ingram (6) as commissioners, and John Charlton (9 parliamentary awards as commissioner and 5 as surveyor) as its surveyor.
10. J. Chapman, 'The Common Lands of Portsea Island ', *Portsmouth Papers*, 29 (1979).
11. North Yorkshire R.O., West Tanfield enclosure papers.
12. J. Chapman and T.M. Harris, 'The Accuracy of Enclosure Estimates: some evidence from Northern England', *Journal of Historical Geography*, 8 (1982), 261-4.
13. Including those under the 1836 General Act, which had no individual act.
14. See M.E. Turner (ed.), *A Domesday of English Enclosure Acts and Awards by W.E. Tate* (Reading, 1978), for a full list of missing estimates.
15. Dorset R.O., Enc. 34; P.R.O., MPB 6.

16. M.R. Postgate, 'Field Systems of Breckland', *Agricultural History Review*, 10 (1963), 99-100.
17. See J. Chapman, 'Efficiency in Land Redistribution: the case of the English enclosure movement', in J. Vervloet and A. Verhulst (eds.), *Papers presented to the Permanent Conference for the Study of the European Rural Landscape* (forthcoming, 1991).
18. For an example of the former see Charlton Marshall, Dorset R.O., Book iv, pp 490ff.   The latter is specifically recorded in the Chesterfield award, Derbyshire R.O., A86.
19. For example, Arlingham, Gloucestershire. Gloucestershire R.O., Q/RI 7.
20. Gloucestershire R.O., Q/RI 13
21. Somerset R.O., Q/RDe 70.
22. P.R.O., MPE 120.
23. Dorset R.O., D396.
24. Gloucestershire R.O., Q/RI 24.
25. P.R.O., MR 615.
26. Somerset R.O., Q/RDE 119.
27. For example, Compton Martin, Somerset R.O., Q/RDE 71, gives full details only of the purchasers.   For two opposing views on who lost see J. Martin, 'The Small Landowner and Parliamentary Enclosure in Warwickshire', *Economic History Review*, 2nd Series, 32 (1979), 328-43, and G.E. Mingay, *Enclosure and the Small Farmer in the Age of the Industrial Revolution* (1968).
For the impact of public sales see J. Chapman, 'Land Purchasers at Enclosure: evidence from West Sussex', *Local Historian*, 12 (1977), 337-41.
28. Gloucestershire R.O., Q/RI 81.
29. For example, at Sawston, Cambridgeshire, the commissioners objected to the way the clerk had drawn up part of the award, and ordered it to be 'altered to a plan recommended by the Commissioners'.   Cambridgeshire R.O., Add MSS 6065.
30. For example, J. Crowther, *Enclosure Commissioners and Surveyors of the East Riding* (East Yorkshire Local History Society, 1986).
31. The bill for Upper Snodsbury, Worcester, was printed with blanks for the names of the commissioners, but with 'gentleman' after each.
32. Not to be confused with the commissioners who carried out the detailed work in the field for the earlier awards.   Their nearest equivalent after 1845 were the 'valuers'.

CHAPTER FIVE

The Tithe Surveys of South-West England

Roger Kain, Richard Oliver and Jennifer Baker

This chapter is arranged in three sections: the first discusses tithe surveys in general and the evidence they contain on past landscapes, societies and economies; the second highlights the particular characteristics of the tithe maps of Cornwall, Devon, Somerset and Dorset; and the third illustrates the varied form and content of tithe maps by reference to some illustrations from Dorset.[1]

1. Tithe Surveys and their Evidence

Tithe surveys are among the most important manuscript sources used by historians researching questions concerning land ownership, management and use, while to local historians investigating the history of particular places, tithe maps provide a large-scale picture of a parish as it was some 150 years ago. In the Public Record Office, 'tithe maps and related records continue to be the single most popular category of cartographic record for written enquiries and personal searches'.[2]

Tithe surveys originated as a result of the Tithe Commutation Act of 1836 which reformed the way in which the established church was financed by a tax (the tithe) on local agricultural output. A parish tithe survey consists of three related documents: *tithe apportionments* are the legal instruments which specify the amount of the reformed tax (tithe rent-charge) apportioned to the owners of particular land parcels (tithe areas), *tithe maps* identify these tithe areas (usually individual enclosed fields, strips in open-field parishes, houses, gardens etc.) and provide a record of their boundaries, and *tithe files* contain the locally-generated papers from the process of tithe commutation such as minutes of meetings between tithe owners and tithe payers and reports from assistant tithe commissioners or local tithe agents.

It is not clear when the Church began collecting tithes, the traditional tenth of a farmer's produce given to support the Church, but by the nineteenth century there was certainly great confusion about the manner of payment. This was made locally by each farmer giving to support his own priest. Over the centuries, the peculiarities, ambiguities and irregularities of local custom multiplied. Precedent was piled upon precedent and fresh complications were caused by the dissolution of the monasteries and the enclosure of open fields. By the beginning of the nineteenth century it was no longer possible to discern even the vaguest outlines of a system amongst a host of local practices and customary arrangements. The courts sanctioned both gross oppressions by the Church and flagrant evasions by tithe payers. Discontent over tithe payment erupted into violence before a remedy was found in the Tithe Commutation Act of 1836. The essence of the settlement was the substitution of a money payment, fluctuating from year to year in accordance with the price of wheat, barley and oats, for all customary payments whether in kind (as, for example, sheafs from the corn field, hay from the meadows, apples from the orchard, piglets from the litter) or in cash. This rent-charge was fixed initially according to the actual value of tithes collected in a tithe district (usually a parish in southern England and the Midlands, and a township in the North) and was apportioned among its farmers according to the type and value of land which they occupied. All of this necessitated a detailed field survey of parishes from which tithes were payable.[3]

The rural landscape of England and Wales and the townscapes of some urban settlements are depicted exactly in these field-by-field tithe surveys. The enquiries of the Tithe Commissioners covered that three-quarters of the country where some tithe remained payable in 1836. In terms of coverage the four south-western counties stand right at the head: all the 212 tithe districts of Cornwall possess tithe surveys (though some individual landed estates were exempt from tithe), 466 of the 478 Devon districts, 270 of the 283 in Dorset, and 482 of the 501 in Somerset have extant tithe surveys (see Table 5.1). Overall in terms of parishes the coverage amounts to some 97 per cent.

The maps drawn for each parish show the boundaries of fields, woods, roads, streams, the position of buildings and often much else of local topographical interest (see section 2 below), while the accompanying schedules of apportionment give the names of the owners and occupiers of each tithe area, describe and name the property, and list the state of cultivation (land use) and statute acreage of each tithe area. Parish tithe files often contain descriptions of local landscapes and farming practices and statistics on crops, yields and livestock numbers. In total, the amount of information which the tithe surveys can provide about land tenure, field systems, land use and farming is unequalled by any other series of documents. Their accuracy is sufficient to warrant their continued use in courts of law and their comprehensiveness and uniformity is surpassed only by such as the Land Utilisation Survey of Great Britain

Table 5.1 Some Summary Characteristics of South-Western Tithe Maps

	Cornwall	*Devon*	*Dorset*	*Somerset*
Total number of tithe districts	212	478	283	501
Total number of tithe maps	212	466	270	482
Printed apportionments	103	80	2	19
Districts apportioned by holding rather than by field	2	8	15	11
Maps at 3-chain (1:2376) scale	103	274	43	119
Maps at 4-chain (1:3168) scale	85	105	20	65
Maps at 6-chain (1:4752) scale	8	61	134	201
Maps at 8-chain (1:6336) scale	4	9	34	49
First class maps	72	93	3	50
Maps with construction lines	94	179	7	53
First class maps with construction lines	58	79	3	42
Printed map	0	4	6	40
Maps on which the mapmaker/ surveyor is named	180	397	156	252
Maps showing turnpike roads	27	28	16	18
Maps showing industrial land use	43	20	3	18
Maps showing woods	86	208	149	219
First class maps showing woods	26	28	0	19
Maps showing plantations	51	147	81	144
Maps showing woods and plantations	46	140	74	134
Maps showing parkland	46	68	39	36
Maps showing arable and grass land	2	14	51	32
First class maps showing arable and grass land	0	0	0	0
Maps showing gardens	10	25	8	42
Maps showing farmyards	29	64	5	20
Maps showing open-strip fields	10	3	31	27
Maps showing hedges	9	41	22	48
Maps showing fences	22	126	55	88
Maps showing both hedges and fences	3	19	10	26
Maps showing hedge ownership	3	28	15	44
Maps showing fence ownership	3	1	0	34
Maps showing field gates	23	82	43	6
Maps with land ownership info.	8	11	14	18
Maps with field names	3	10	30	17
Maps with cartouches	14	21	1	5
Maps with decorative elements	15	21	1	6

undertaken in the 1930s. Indeed, they rank as the most complete record of the agrarian landscape at any period.

(i) Tithe maps as historical sources

One of the most important of the tithe survey documents is the map; the characteristics of those of the South West are discussed in section 2 below (also summarised in Table 5.1). The recommended scale for tithe maps was one inch to 3 chains or about 26.7 inches to a mile but many large parishes were surveyed at smaller 4 or 6 chain scales. The technical specifications for tithe maps were drawn up by Lieutenant Robert Kearsley Dawson of the Royal Engineers who was seconded to the Tithe Commission in 1836. He very much hoped that tithe maps might be assembled together and published as a General Survey of the whole country but for financial reasons the government decided against this.[4] It would have been very expensive to resurvey areas where land owners already possessed estate maps of sufficient accuracy to be incorporated into tithe maps.

Most tithe maps of open-field parishes, like that of Winterborne Kingston in Dorset (Figure 5.1) show strip fields by dotted lines and closes by solid lines. It is thus possible to reconstruct the pattern of open field from tithe maps. In a few parishes, parcels of land have passed undivided from generation to generation, but these places are exceptions to the general rule. Over much of the country changes in ownership have been frequent, estates have been broken up and land has been put to new uses. Altered tithe apportionments and their maps record major changes in the shape, size and status of the original tithe areas resulting from severance or subdivision. The building of railways, construction of new roads, re-allocation of land under an enclosure award, and other public works all necessitated altered apportionments. When changes affected a large part of a parish, a whole new apportionment might be made. In 1848 the parish of Winterborne Kingston was enclosed and its pattern of fields, land ownership and land tenure totally transformed. Such was the degree of alteration that a complete re-apportionment of the rent-charge was needed (Figure 5.2); comparison of the original map and this altered apportionment map provides fascinating glimpses of the rural landscape before and after enclosure.

As is apparent from section 2 below, there is considerable variation in the amount of other detail shown on tithe maps. A few of them use a system of conventional symbols recommended by Dawson to identify different types of land and these can be read as virtual land use maps (Figure 5.3). Characteristically inhabited buildings are tinted red on tithe maps and barns and other such structures in grey. Tithe maps of industrial or mining districts often depict these enterprises in great detail (Figure

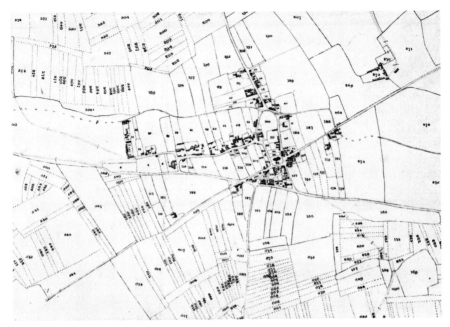

Fig.5.1 Part of the tithe map of Winterborne Kingston, Dorset in 1844 showing the presence of open fields at this date. Photograph reproduced by permission of the Keeper, Public Record Office.

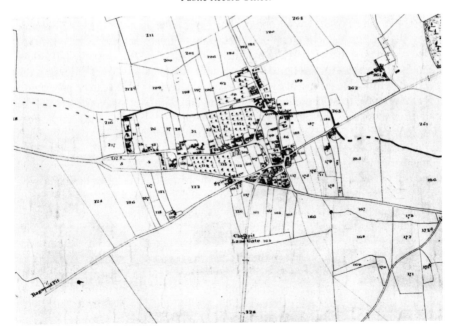

Fig.5.2 An extract from the altered apportionment tithe map of the same part of Winterborne Kingston made in 1848 after enclosure. Photograph reproduced by permission of the Keeper, Public Record Office.

5.4) and Figure 5.5 illustrates the kind of detail with which urban morphology is sometimes portrayed.

All tithe maps were tested for accuracy in Lieutenant Dawson's office in the Tower of London and maps which passed all his rigorous checks are known as 'first class maps'. Others which failed on some count (usually because they were drawn at a scale smaller than one inch to 4 chains) but which were nevertheless sufficiently accurate for the immediate purposes of tithe commutation are known as 'second class maps'. First class maps can be identified by the presence of the Tithe Commissioners' official seal and a signed certification of map accuracy. The comparative value of first and second class tithe maps as historical sources is discussed further in section 2 below.

(ii) Tithe apportionments as historical sources

A tithe map used on its own can provide only a limited amount of information about rural landscapes, society and economy. Much more can be obtained if the map is interpreted in conjunction with the apportionment roll. Each tithe area on the map is assigned a reference number and against this number in the apportionment more information is given.

First, the names of the owner and occupier of a tithe area are stated. These are very valuable data and in this country, which lacks a full cadastral (property) survey on the continental model, they are unique to tithe surveys. It is possible to identify all the fields owned or occupied by particular individuals and by piecing these together, maps of the pattern of ownership and occupation of land can be produced.[5] It is also possible to work out the size of estates and farms as the sixth column of the apportionment states the surveyed acreage of each field. Such studies can be very time consuming if carried out over a large area but are made much easier by simple computerised data processing.[6]

In parishes where the solid lines on tithe maps indicate that fields were enclosed at the time of tithe commutation, the pattern of the field boundaries has been used to diagnose earlier tenurial conditions. Arguing in this way from form as represented on maps to hypothesised functional arrangements at earlier dates can be dangerous but if tithe apportionment evidence of land tenure is used also, deductions become more certain. For example, inspection of the extract of the Otterton, Devon, tithe map reproduced in Figure 5.6 indicates the existence of a number of fields of elongated, strip-like shape which suggests that they might be relics of an open-field farming system from before the nineteenth century. Reference to the tithe apportionment (Figure 5.7) reveals that the small block of fields numbered 671, 672, 673, 710, 711 and 712 in the extreme north of this map extract were all part of the estate

Fig.5.3   The 1838 tithe map of Gittisham parish in Devon differentiates land use by colour washes and represents woodlands, orchards etc. by conventional symbols.   Photograph reproduced by courtesy of Devon Record Office.

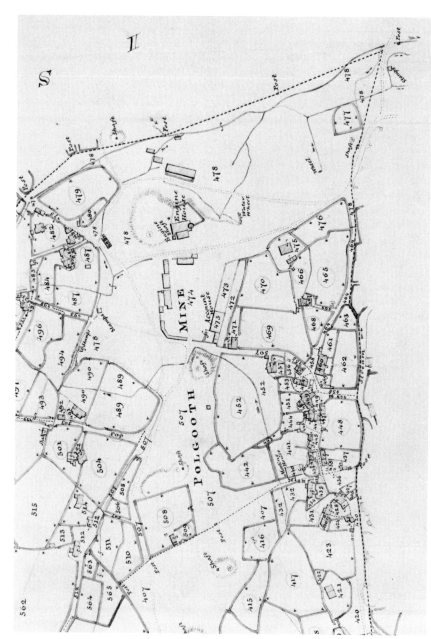

Fig.5.4   An extract from the tithe map of St. Mewan, Cornwall showing the Polgooth Mine workings.   Photograph reproduced by courtesy of Cornwall Record Office.

Fig.5.5   Part of the tithe map of Stoke Damerel parish in Devon with details of the docks and town of Devonport.   Photograph reproduced by courtesy of Devon Record Office.

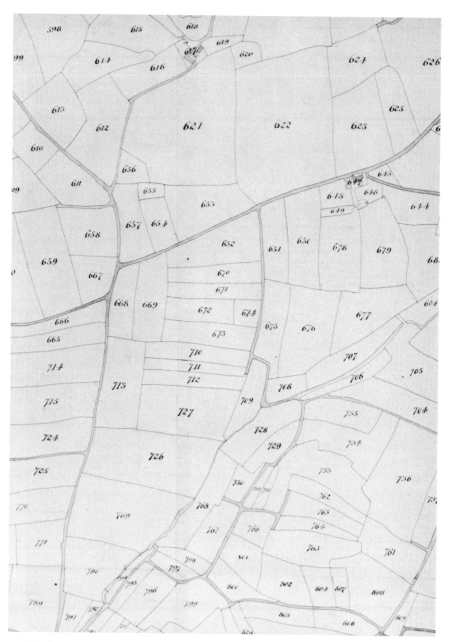

Fig.5.6   Part of the tithe map of Otterton, Devon made in 1844. Photograph reproduced by
courtesy of Devon Record Office.

of Lord Rolle at the time of tithe commutation but that each field was held by a different occupier and thus each formed part of six separate and still spatially fragmented farm holdings. The form of Otterton village itself suggests that in the medieval period it was the focus of an open field agrarian system. In 1844 at the time of the tithe survey as today, farmhouses fronted on to the main street with buildings strung out behind them on long narrow tofts with access from a back lane. Though tithe surveys were produced late in the agrarian histories of places such as Otterton which was enclosed for sheep farming in the Tudor period, they do provide a picture before all the more modern and especially post-World War Two changes that have taken place in the countryside. With the tithe apportionment to supplement a tithe map, the tithe survey can be much more useful than say an equivalent large-scale Ordnance Survey map as a starting point for retrospective enquiries in landscape history.[7]

Each tithe area is also named and described in the apportionment; for students of place, field, property and other local names, the sheer number of these data in tithe surveys is unequalled by any other historical source. As well as being of interest in their own right, the property names of tithe apportionments can be used in conjunction with the tithe maps to link households enumerated in the 1841 or 1851 population censuses to the houses in which they resided on census night.[8] Once houses have been 'repopulated' the residential pattern can be analysed.

Finally, the land use of each field is recorded in the tithe apportionment. Tithe surveys distinguish between arable, pasture and wood and often identify land such as marshes, meadows, downs, moorlands, hop grounds, gardens, orchards, fruit grounds and unfarmed property. This information used in conjunction with the tithe maps enables maps of mid-nineteenth-century land use to be reconstructed.

(iii) Tithe files and their evidence of past farming systems

One of the problems with tithe map and apportionment land use information is the very crude classification used by the surveyors and valuers. Only very rarely are the actual crops grown recorded in the apportionments or is any idea of particular agricultural practices provided. No South-West England tithe apportionments record crops. Tithe maps and apportionments may provide a reliable record of generalised land use and of land ownership and occupation, but they say little about the reasons for the particular land use or ownership patterns which they reveal. They present an essentially static picture; explanations must be sought in other sources, notably estate records. However, the third category of tithe survey documents, the tithe files, do help in this regard because in addition to the official record of the process of tithe commutation in a parish which they contain, very many also preserve

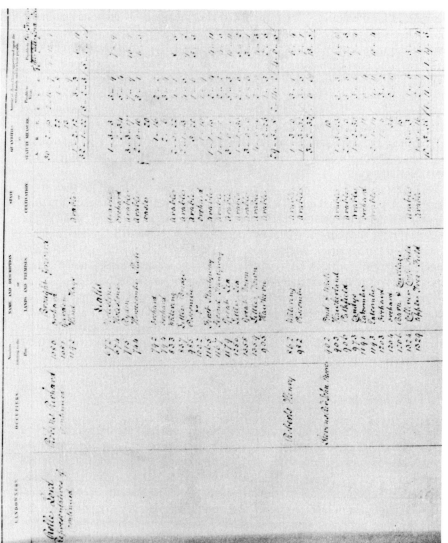

Fig.5.7   A page from the tithe apportionment of Otterton, Devon.   Photograph reproduced by courtesy of Devon Record Office.

printed questionnaires which were completed by the assistant tithe commissioner or local tithe agent who officiated at the parish tithe commutation.[9] The information recorded in these questionnaires was used to assess whether a voluntary agreement for tithe commutation really was fair to all parties. In the country as a whole tithe was commuted by voluntary agreement in a little more than half the tithe districts; in Somerset and Dorset it was two-thirds, in Cornwall almost exactly a half, and in Devon a little less than 50 per cent.[10] The questionnaires contain descriptions of agricultural practices which can give valuable insights into the working of farms in the mid-nineteenth century. In addition they record the acreage and yield of crops grown on the parish arable and for western and south-western counties there are also rough estimates of the numbers and type of livestock pastured on the grasslands.[11]

2. The Tithe Maps of South-West England

Although the tithe surveys of *c*. 1837-52 covered all parts of England and Wales and were derived from the same national legislation and organised by one centralised Tithe Commission in London, there are pronounced variations in their characteristics and contents between regions and, indeed, between counties. The South West is no exception to this.

Tithe surveys were produced by individual surveyors working to the barest minimum specification. That barest legal minimum was to show the boundaries of tithe areas. Very occasionally tithe areas were whole holdings or estates rather than the usual individual closes or open-field strips; Dorset was exceptional in this regard, with as many as 6 per cent of tithe districts with rent-charge apportioned by holdings. There was no legal requirement to show more than these details of tithe areas, and, indeed, after difficulties arising from the drafting of the original Tithe Commutation Act had been resolved in 1837, land owners could submit to the Tithe Commissioners any map with which they, the land owners, were satisfied. In practice, tithe surveys cover the whole spectrum from minimalist maps showing only tithe areas, to *de luxe* productions outdoing in their portrayal of topographic detail even the early Ordnance Survey 1:2500 plans.   The great majority of tithe maps fall between these two extremes and there are three reasons which may explain this. First, the maps were produced by numerous individual surveyors, some of whom might map up to 10 or 15 per cent of a county, but many of whom produced only one or two maps, and all of whom had their own ideas as to what to show. Second, many tithe maps were copies of earlier maps, and third, the maps were commissioned by land owners who would no doubt have had their own ideas as to map content. Bearing in mind that the maps had to be paid for by land owners as part of a compulsory process which could be expensive even without any 'frills', the remarkable thing is not that the maps as a whole are so much less 'finished' than were the Ordnance Survey maps of a few decades later, but that they contain so much more detail beyond that which was strictly necessary for their immediate purpose.

(i) First and second class tithe maps

The precise number of first class tithe maps has not, as yet, been accurately determined; for England and Wales as a whole it is probably about 12 per cent by number, but not more than 7-8 per cent of the country by area.[12] The South West as a whole conforms to this (208 of the 1430 maps are first class), but there are pronounced variations between counties: Cornwall has 34 per cent, Devon 20 per cent, Somerset 10 per cent, but Dorset only 1 per cent.

The attitude of the modern user towards the first class maps is likely to be very different from that of contemporaries; first class tithe maps maps say more about the context of their making than about the landscape itself. Contemporary critics saw them as the only redeeming feature of a generally unsatisfactory operation and used the comparatively small number of first class maps to argue for a complete large-scale re-mapping by the Ordnance Survey.[13] Although no doubt planimetrically more accurate than most of their second class brethren, first class maps often contain *far less* information of use to the present-day historian. Babcary in Somerset is a good example of the most austere type of first class map, with no colour and buildings shown in outline only. Other first class maps do shade buildings and distinguish between houses and uninhabited buildings, but first class maps which show much detail about land use are very unusual; not a single one in the South West shows details of arable and grassland, and only a third of first class maps show woodland by symbols as against nearly half of the second class maps.[14] In the debate about the nature of tithe mapping in 1836-7 the Tithe Commissioners made much of the advantages of having the boundaries of tithe areas accurately mapped on first class maps so that future disputes as to liability to tithe could be avoided but in practice most land owners were apparently willing to take the risk of not having such a record. The vast majority of first class maps would have been specially made, rather than re-plotted from existing fieldbooks, and it is possible that at least some first class maps were commissioned by land owners in tithe districts for which there were no suitable maps at all, but who felt that while they were about it they might as well do, in local parlance, 'a proper job'. There is also a tendency for first class maps to be concentrated in certain areas; for example, most of those in Somerset are in the north-eastern half of the county, though the reasons for such patterns are not entirely clear.

From the point of view of contemporary users who demanded the strict planimetric accuracy exemplified by the Ordnance Survey, the small proportion of first class maps was enough to condemn the tithe surveys as a whole but such perjorative comments do not really devalue tithe maps as historical sources. Field boundaries are not always in their strict places but the vast majority of fields shown on the tithe surveys were still in existence some forty or fifty years later when they were recorded to 'first class' standards by the Ordnance Survey and so the pattern represented on tithe maps can be carried through to the present by comparative examination of tithe and Ordnance Survey maps. Tithe survey information can be transferred to modern base maps without much difficulty, especially in South-West England where the field pattern has been less altered than in some other parts of the country.

It is probable that not all maps which were intended to be first class were accepted as such by the Tithe Commissioners: in any county there are second class maps which contain the original construction lines, a perceived *desideratum* for achieving first class status. Indeed in Devon, there are more second class maps with construction lines than there are first class maps.[15] First class maps are sometimes found without construction lines as three copies were made of each map and usually only one will bear the evidence of construction. One copy was for the Tithe Commissioners (most of these are now held in the Public Record Office at Kew), one was for the Diocesan Registry (these now form the basis of county record office sets of tithe maps), and one was for the parish (often no longer extant but occasionally still with local churchwardens, or perhaps transferred to county archives). The Tithe Commissioners usually kept the original but evidently not always and a proportion of original maps, the extent of which can probably never be ascertained, have found their way into county record office collections of tithe maps.

Although the number of tithe maps explicitly admitted to be copies of earlier mapping is small, both in the South West and elsewhere, it is a reasonable surmise that the majority of second class maps were such copies, either of whole tithe districts or of estate maps. Legally, land owners were only obliged to map tithable land, but in practice tithe-free land was usually mapped as well. This would have been an inexpensive task if existing mapping was available for copying; where tithe-free and tithable land was intermixed it might well have been more cost-effective to map the whole. Another clue to the copying of earlier mapping is where the same prolific map-maker uses a variety of scales without apparent regard to terrain or field size.[16]

Tithe maps usually remained in manuscript, but sometimes they were lithographed. Lithography had the advantage that only a manuscript original and a tracing of it needed to be drawn and as many copies as were needed could then be printed from stone. Whether many copies above those wanted for immediate tithe commutation purposes were actually produced for sale remains unknown. Most lithographed tithe maps were produced by Standidge and Company of London, a firm which did much official work for H.M. Stationery Office. The proportion of lithographed tithe maps in any one county varies: there are none at all in Cornwall, but as many as 8 per cent of all Somerset tithe maps are lithographed. The apportionments were sometimes printed as well; nearly half of those in Cornwall were treated thus, but less than one per cent of those in Dorset. Again, the reasons for the great variation in practice are far from clear.

(ii) Scales

The scales of tithe maps show considerable variations between counties, individual districts and surveyors. They are usually in the range of 1 inch to 3 chains (1:2376) to 1 inch to 12 chains (1:9504); smaller scales are occasionally met with in upland Britain, and larger scales in urban areas. The Tithe Commissioners would not consider testing for sealing as first class any map at smaller than the 4 chain (1:3168) scale, and the majority of first class maps were at the 3 chain scale. The 3 chain and 6 chain scales (1:4752) were the most commonly used, though there are pronounced regional variations: in Cornwall about 46 per cent of maps are at 3 chain and about 40 per cent at 4 chain scales and in Devon the figures are about 58 per cent and 22 per cent respectively. In Dorset only 15 per cent are at 3 chains to an inch and 7 per cent are at the 4 chain scale; in Somerset the comparative figures are 24 and 12 per cent. What is clear is that there is no correlation between the proportion of a county mapped at the first class scales and its coverage by first class maps.

It is common for tithe maps to have separate or inset enlargements of detail, usually of settlements or parts of settlements, and sometimes of open fields. A peculiarity of some Somerset maps is the exceptional number of such enlargements: an extreme case (nationally as well as regionally) is the map of Wivelscombe where the main part is at 6 chains but is supplemented by no fewer than 41 enlargements of detail around the margins of the map. At Midsomer Norton there are 29 enlargements of detail. Such bizarre maps prompt the question of whether a much larger scale should have been used from the outset!

(iii) Surveyors

Surveyors and map-makers fall into three categories: those who worked locally and produced a large number of maps, those who worked locally and produced only a few maps, and those who came from some distance away, from outside the county, or even from London. Good examples of the first category are the Woodmass partnership based at Chudleigh in Devon who worked in Devon, Cornwall and Somerset, H. and R. Badcock of Launceston who worked in Devon and Cornwall,

Thomas Oates Bennett of Bruton who worked mainly in Somerset, James Poole of Sherborne who worked mainly in Somerset and Dorset, and John Baverstock Knight of Piddlehinton, who worked in Dorset. Less prolific, though locally important, are, for example, Thomas Lock of Instow, Charles Cooper of Bideford and George Northcote and Hugh Ballment, both of Barnstaple. Whilst the employment of such men with obviously well-established local practices or of local men willing to undertake valuations and copy existing maps needs no special explanation, the use of men from as far away as London merits further investigation. Various reasons may be surmised: that some of them tendered successfully in response to newspaper advertisements (Figure 5.8), that some regularly worked for absentee land owners or their agents,[17] that some absentee land owners who usually resided in, say, London, or corporate land owners such as Oxford and Cambridge colleges found it easier to recruit a surveyor in London than locally. Many of the local map-makers and surveyors - T.O. Bennett of Bruton is a good example - also acted as valuers and agents. The great growth in demand for tithe surveys from 1837 onwards, and for railway surveys a few years later, led to a corresponding increase in demand for surveyors' services, and it is unsuprising to find that surveyors deserted the Ordnance Survey of Ireland for the much greater rewards of tithe and railway surveying in England.[18]

The actual standard of execution varies from map-maker to map-maker: first class maps are usually either plain and neat, or 'working plans', plain and with linework of variable gauge and rough accompanying writing; James Peachey Williams of Bridgwater was a man who produced both varieties. T.O. Bennett of Bruton habitually produced neat maps on paper, including some first class; William Wadman of Martock equally habitually produced much rougher-looking ones on parchment, often at 8 or 9 chain (1:6336 or 1:7128) scales. Aller in Somerset has a beautifully coloured map by Richard Dixon of London, while just a few miles away that of Wedmore is one of the 'roughest' of all English or Welsh maps. Occasionally an insight is given into the problems of surveying. The map of Holcombe Burnell, Devon, carries a note by the surveyor (J. Philp of Exeter) that 'the land which this map represents is nothing scarcely but Hill and Dale - hence many of the lines will not prove to that degree of accuracy which they would if the ground was level.'[19]

(iv) Non-statutory detail on tithe maps.

Reference was made in section 2 (i) above to two features of the tithe maps: that the amount of detail actually recorded is generally much greater than was strictly necessary for the purposes for which the maps were made, and that the second class maps are generally a much better source for such non-statutory land use and general

hour of Eleven in the Forenoon.
Given under my hand this Sixth day of April, 1840.
THOS. BUCKLER LETHBRIDGE, by W. CROOTE,
his Agent, duly Authorized.

Duty Free.]   **TITHE COMMUTATION ACT.**
TO LAND SURVEYORS AND APPORTIONERS.

PERSONS desirous of CONTRACTING for MAPPING and APPORTIONING the Rent Charge in lieu of Tithes, of the Parish of EXBOURNE, Devon, containing about 1700 Acres, under the Tithe Commutation Act, are requested to send TENDERS for the same, free of expense, to Mr. JOHN TATTERSHALL, the younger, of Exbourne, on or before the 11th day of MAY next.

Persons Tendering for the MAPPING, are. desired to state the price per Acre, at which they will furnish a first class MAP and BOOK OF REFERENCE, with Two Copies thereof, according to the printed instructions of the Commiss'oners, to be subject to their approbation and seal, and also within what time they will engage to perform the work.

Persons Tendering for the APPORTIONMENT of the RENT CHARGE, are desired also to state at what (if any) additional charge, they will at the same time make a VALUATION for the POOR RATE of the said Parish.

Any further particulars may be obtained on application to Mr. TATTERSHALL, either by letter or otherwise.

Dated 1st April, 1840.

TEIGNMOUTH AND SHALDON BRIDGE.

NOTICE IS HEREBY GIVEN, That this Bridge will be Re-opened for Horses and Carriages, on

Fig.5.8   Advertisement placed in *Trewman's Exeter Flying Post,* 16 April 1840, for a surveyor and tithe apportioner at Exbourne, Devon.

topographic information than are the first class maps. The reason for this probably lies more in the habitual practice of surveyors than in any deference to the model set of conventions produced by Lieutenant Robert Kearsley Dawson as part of his abortive attempt to have the tithe maps made to a consistent specification as part of a national cadastral survey.[20] The particular land uses mapped vary from county to county within the South West. Woodland is fairly generally shown, with somewhat under half of the maps in each of the four south-western counties indicating woodland on the face of the map as well as listing woodland parcels in the apportionment. The portrayal of arable and grass shows more pronounced variations, with 0.5 per cent of maps in Cornwall and 3 per cent in Devon recording this information, against 6 per cent in Somerset and over 20 per cent in Dorset. Gardens show a different pattern: they appear on 4 per cent of Cornwall, 5 per cent of Devon, 3 per cent of Dorset but over 8 per cent of Somerset tithe maps. Farmyards are much more often shown distinctively in Devon and Cornwall than in Dorset and Somerset, which must reflect the practice of map-makers rather than differences in the reality of agrarian operations. Land use is usually shown by colour wash, gardens are sometimes shown by patterns of dashes and conventional symbols ('bushes'). The fact that 20 per cent of Cornish and 4 per cent of Devon and Somerset tithe maps show industrial features is proportionate to the relative industrialisation of these counties, but it remains to be investigated as to how many 'mills' or nondescript 'uninhabited' buildings on the maps are listed as industrial rather than agricultural buildings in the apportionments. The detail with which industrial and transport features are shown varies greatly but not more significantly within the South West than in England and Wales as a whole. Industrial premises are usually shown in plan but it is rare for the functions of individual buildings to be given on the face of the maps (they may be given in the apportionments); in that respect the standard of depiction in Cornwall is somewhat better than in England as a whole. It is unusual for more than the outside boundary fences of railways and tramways to be shown. Turnpike roads are usually indicated, if at all, only by the incidental naming of toll houses; contemporary Ordnance Survey and commercial maps are potentially a much better source of information on turnpikes than are tithe maps. Only canals are recorded on tithe maps in any real detail, with locks indicated at least by a narrowing of the channel if not by symbol, and towpaths almost always included.

As a general rule all topographic details are shown in plan on tithe maps, with conventional, non-planimetric symbols reserved for land use. There are exceptions and in South-West England the majority of these are in Somerset where, particularly in the southern half of the county, a number of churches are shown pictorially by somewhat conventionalised small drawings, though care is often taken to show spires for those churches which have them in reality. At Barwick, also in Somerset, strange pictograms appear, which prove on reference to the apportionment to be a pigeon house and a tower, but such pictorialism is unusual on tithe maps. More common in Somerset, where it appears on 9 per cent of maps, is the showing of hedge and fence ownership by pictorial representations, with the hedge or fence incursing into the field to which it belongs. This practice is less common in the other south-western counties.

(v) Decoration

Possibly the most striking feature of south-western maps generally, and particularly of the Devon and especially Cornwall maps, is the amount of incidental decoration which they contain. In the eighteenth century it was common for both small-scale county maps and large-scale estate maps to carry decorative cartouches, sometimes supplemented by rural scenes or allegorical figures. By the time of the tithe surveys such decoration was unusual and tends to be confined to such minor decoration as scale-bars styled and coloured like a wooden ruler. However, in Devon and Cornwall a number of tithe maps carry ornamental scenes which can surely only have been included out of whimsy. To give three examples: Sithney in Cornwall has a small pastoral scene including a cow, a man climbing a tree and two men fishing. On the maps of Northam and Woolfardisworthy in Devon, both drawn by Benjamin Herman of Northam, the coastline is decorated with rowing boats and vessels in full sail and various surveying instruments are arranged around the scale bars. Such gratuitous decoration is quite unknown on the maps of Dorset and Somerset and it is equally unusual on maps elsewhere in England and Wales. Indeed, a very high proportion of the national total of 'decorative' tithe maps are concentrated in Devon and Cornwall.

3. Illustrations of Some Dorset Tithe Maps

This chapter concludes with some illustrations of Dorset tithe maps briefly described to underscore the variety of maps produced in an individual county as a result of the implementation of the Tithe Commutation Act of 1836.[21] Though the tithe surveys derive from this single piece of national legislation, they are no less varied cartographically than, for example, the body of enclosure maps discussed in Chapter 4.

Marsh Caundle[22]

The Marsh Caundle tithe map was produced in 1838 (Figure 5.9). Marsh Caundle was valued and surveyed by James Poole of Sherborne. The map is drawn at a scale of one inch to 6 chains and is highly coloured to indicate property ownership, although individual field boundaries are not given as in this parish tithe areas were equated with land holdings, not fields. The common is named and woodland areas are depicted by rather idiosyncratic symbols.

Nether Compton

The Nether Compton tithe map was made in 1839 (Figure 5.10). Nether Compton was valued by Edward Watts of Yeovil, Somerset, and Edward Thomas Percy of Sherborne, Dorset and was surveyed by Edward Watts. The map is extensively annotated and carefully drawn and is also at the 6 chain scale. The glebe lands are indicated on the map, and other features of local topography include a lime kiln, a lodge and a water mill.

Bridport

The Bridport tithe map was made in 1845 (Figure 5.11) and is an example of those produced from within the locality. Bridport was valued by a local man - Richard Cornick - and surveyed by R. James of nearby Chideock. The map is drawn at a scale of 3 chains to an inch.

Leigh

The tithe map and apportionment of Leigh are dated 1840. The map shows both the irregularly shaped early enclosures that were made in Leigh (Figure 5.12) and the more rectangular later enclosures (Figure 5.13).[23]

West Stour

At West Stour, in the Blackmore Vale (Figure 5.14) there is surviving evidence of the nature of the commission awarded to the valuer and surveyor. The valuers, John Raymond of Shaftesbury and Thomas Davis of Warminster, were required to determine what would be an equitable annual payment to be made to the impropriator in lieu of the 'Corn, Grain and Pulse arising within the said parish', and what sum in lieu of 'all other Tithes arising from all the Lands' should be paid to the incumbent, and to his 'heirs and assigns'. Provisional agreement was reached at a parish meeting in May 1841; this was confirmed by the Tithe Commissioners in May 1842, and signed and sealed by them in May 1843. Payment of tithes in kind ceased in October 1842 and payment of the rent-charge in lieu was recoverable six months later. The tithe apportionment for West Stour reveals that all of the 1014 acres of land in the parish were tithable and that about two thirds of the land (685 acres) was meadow or pasture land, a proportion to be expected in this area of heavy soils. As the tithes of West Stour were commuted by agreement the tithe file contains a printed questionnaire which was used by the local tithe agent to judge the fairness or otherwise of this agreement. His report tells us that the 300 acres of arable land was cultivated on a four-course rotation, with the land divided almost equally into 76 acres of wheat giving a yield of 25 bushels an acre, 76 acres of barley yielding 25 bushels an acre, 75 acres of clover and seeds yielding 22 bushels an acre, and 76 acres of fallow. Such figures suggest that local tithe agents simply determined the usual arable rotation and then calculated acreages of individual crops by relating the rotation sequences to the total arable acreage, in this case by dividing 300 by 4. Land rentals were high at West Stour: for arable land the average rental was 25 shillings per acre and for pasture the average rental was 33 shillings per acre. The following animals were recorded (doubtless rough estimates on the part of the local tithe agent by virtue of the round numbers): 120 cows, 40 bullocks, 600 sheep.

The tithe map of a parish such as West Stour is useful to the historian as it provides a large-scale reference to local topography in the middle of the last century. The value of the map is increased if it is interpreted in conjunction with the tithe apportionment with its additional information on land ownership, land occupation, land use and place names. If the evidence of tithe files is married to the information to be gleaned from the tithe map and tithe apportionment, then a truly detailed and vivid period picture of landscape, society and economy in the early years of Queen Victoria's reign can be reconstructed. Historians of no other region in England and Wales are as generously endowed with these parish tithe surveys as those who research the history of South-West England.

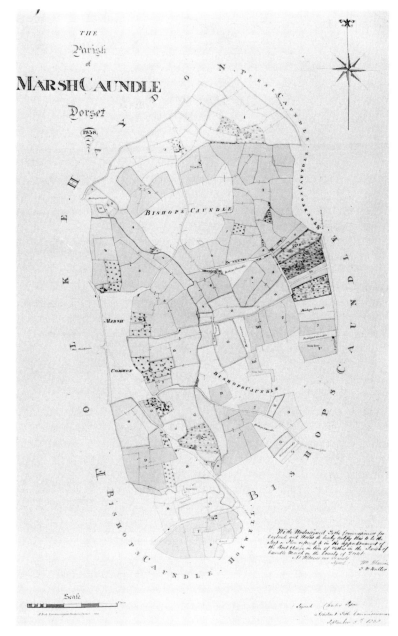

Fig.5.9   Tithe map of Marsh Caundle, Dorset.   Reproduced by permission of Dorset County Record Office; reference T/CDM.

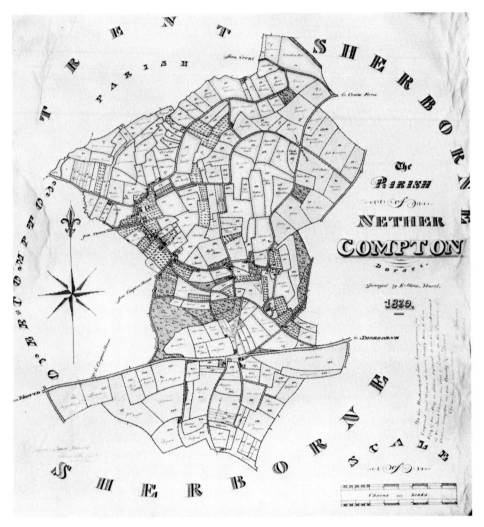

Fig.5.10   Tithe map of Nether Compton, Dorset. Reproduced by permission of Dorset
County Record Office; reference T/NCO.

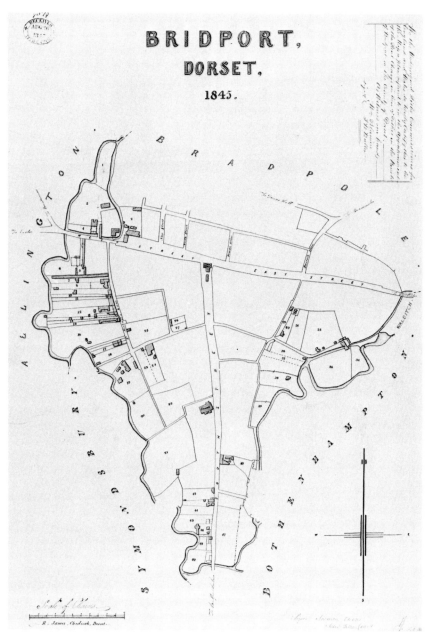

Fig.5.11   Tithe map of Bridport, Dorset. Reproduced by permission of Dorset County Record Office; reference T/BT.

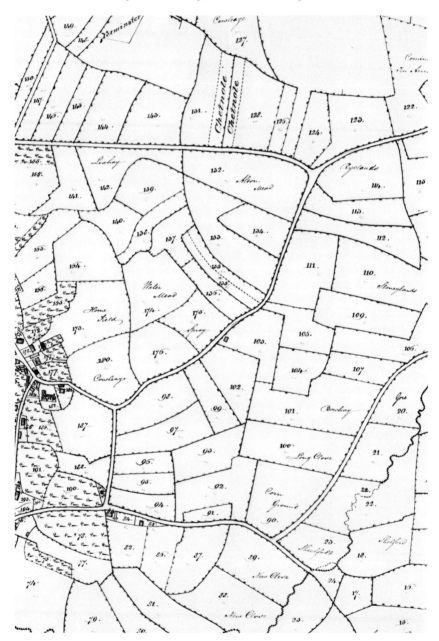

Fig.5.12   Early enclosures on the tithe map of Leigh, Dorset. Reproduced by permission of
Dorset County Record Office; reference T/LEI.

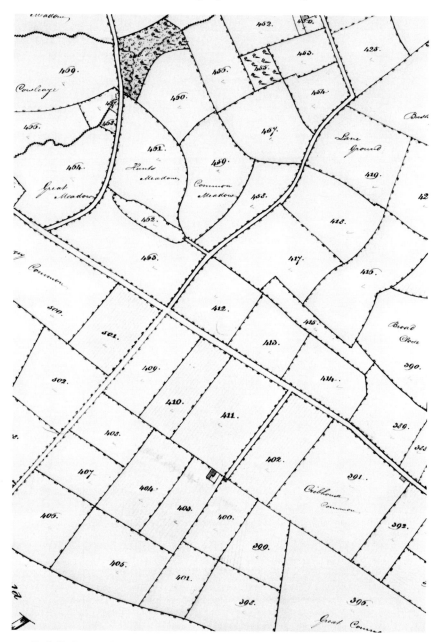

Fig.5.13   Later enclosures on the tithe map of Leigh, Dorset. Reproduced by permission of
Dorset County Record Office; reference T/LEI.

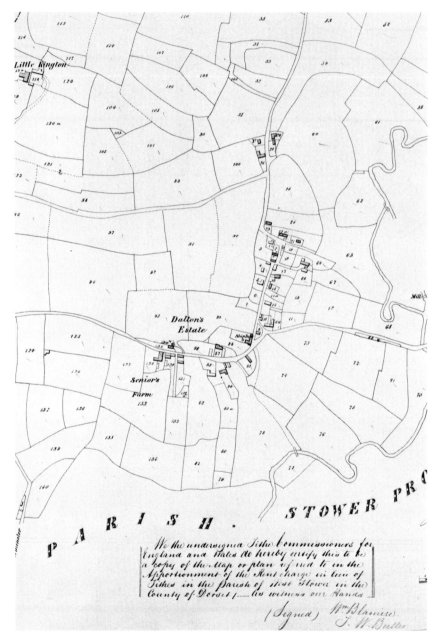

Fig.5.14 Tithe map of West Stour, Dorset. Reproduced by permission of Dorset County
Record Office; reference T/WSR.

Notes and References

1. Section 1 is based on material in R.J.P. Kain and H.C. Prince, *The Tithe Surveys of England and Wales* (Cambridge, 1985); section 2 analyses data obtained from a project funded by The Leverhulme Trust in the Department of Geography, University of Exeter to index and analyse all the tithe surveys of England and Wales; examples discussed in the third section are from J. Gambier (now J. Baker) 'Tithes, Tithe Commutation and Agricultural Improvement: a case study of Dorset, *circa* 1700-1850', (unpublished PhD thesis, University of Exeter, 1990).

2. *The Thirtieth Report of the Keeper of Public Records on the Work of the Pub lic Record Office* (1989), 14.

3. This and the following paragraph draw on R.J.P. Kain, 'Tithe Surveys and the Rural Landscape of England and Wales', *Bulletin of the Society of University Cartographers*, 11 (1976), 1-3.

4. R.J.P. Kain, 'R.K. Dawson's Proposals in 1836 for a Cadastral Survey of England and Wales', *Cartographic Journal*, 12 (1975), 81-8.

5. R.J.P. Kain, 'Tithe Surveys and the Study of Land Occupation', *The Local Historian*, 12 (1976), 88-92.

6. R.J.P. Kain, 'Tithe Surveys and Land Ownership', *Journal of Historical Geography*, 1 (1975), 39-48; 'Extending the Agenda of Historical Inquiry: computer processing tithe survey data', *History and Computing*, 3 (1991), 23-35.

7. R.J.P. Kain and R. Allison, 'Problems and Sources in Historical Geography: agriculture in the past', *Teaching Geography*, 11 (1986), 58-63.

8. The pioneering study of 'house repopulation' was conducted by A. Henstock, 'Group Projects in Local History - house repopulation in the mid-nineteenth century', *Bulletin of Local History, East Midlands Region*, 6 (1971), 11-20. This was reprinted with minor amendments in *Local Population Studies*, 10 (1973), 37-52.

9. R.J.P. Kain, 'The Tithe Files of England and Wales' in M. Reed (ed.), *Discovering Past Landscapes* (1984), 56-84.

10. R.J.P. Kain, *An Atlas and Index of the Tithe Files of Mid-Nineteenth-Century England and Wales* (Cambridge, 1986), 182-230.

10. R.J.P. Kain and H.M.E. Holt, 'Agriculture and Land Use in Cornwall *circa* 1840', *Southern History,* 3 (1981), 139-81.

12. This is an impressionistic figure gathered whilst working on the set of these maps now in class IR 30 at the Public Record Office.

13. See Richard Oliver, 'The Ordnance Survey in Great Britain 1835-1870', (unpublished D.Phil. thesis, University of Sussex, 1986), esp. chapter 5.

14. These and other descriptive statistics of South-Western tithe maps are taken from a computer database of all 11,783 tithe maps of England and Wales preserved in the Public Record Office that is being compiled in the Department of Geography, University of Exeter. See note 1 above.
15. It must be admitted that Devon is exceptional in having so many second-class maps with construction lines.
16. E.g. William Wadman, of Martock, Somerset.
17. There is some evidence of this in Lincolnshire, when the employment of 'outside' surveyors often shows a close relationship with the pattern of landownership.
18. J.H. Andrews, *A paper landscape: the Ordnance Survey in nineteenth-century Ireland*, (Oxford, 1975). One such surveyor was Bland Hood Galland, who in 1834 was a draughtsman in Ireland but a few years later was producing tithe maps in England.
19. P.R.O. IR 30 9/212.
20. Kain, (above, note 4).
21. For further examples, see J. Gambier (above note 1), chapter 5.
22. now Caundle Marsh
23. Dorset R.O. T/LEI. The sections of the Leigh tithe maps illustrated in Figures 5.11 and 5.12 and the enclosure history of the parish are discussed in B. Boswell, *Leigh, A Dorset Village* (1986), 161-4.

CHAPTER SIX

The Ordnance Survey in South-West England

Richard Oliver

The Mapping of the South West and the Origins of the Ordnance Survey

It is appropriate that this essay is to be published in 1991, for this year is being celebrated as the bicentenary of the Ordnance Survey, which though a national organisation nonetheless has particularly notable associations with South-West England. A survey of its activities in the South West touches on many themes in the larger history of the Ordnance Survey, and, indeed, a number of important innovations and developments first saw the light of day on maps of the South West.

It is axiomatic that the Ordnance Survey 'began with the army', which is broadly true: in fact, it started its career as a part of the Board of Ordnance, which, though military, was independent of the Army proper until 1855[1]. The Ordnance was responsible for artillery, fortifications and engineering, all of which in some degree implied surveying and map-making. This was a two-way traffic, since on the one hand the Ordnance needed the services of surveyors and draftsmen, and on the other developed skills which might have an application in civil as well as military affairs. This is epitomised in the person of General William Roy, who, though he died in 1790, is often seen as 'the founder of the Ordnance Survey'; at the beginning of his career, in 1748-55, he was largely responsible for executing a survey of the Scottish mainland at a scale of one inch to 1000 yards (1:36,000), and at the end of it, in 1784-88, he was responsible for carrying out the British part of an operation whereby the observatories of London and Paris were connected by a framework of triangles.

The immediate object of this exercise was to settle certain Anglo-French scientific disputes, and it was carried out under the auspices of the Royal Societies of London and Paris, but Roy was the obvious choice for conducting the British end of the operation, and his subordinates were soldiers. Following Roy's death the unexpected availability of an unwanted, yet very superior, theodolite provided an excuse for the Duke of Richmond to put the trigonometrical work on a permanent basis in 1791, and this is usually taken to be 'the' foundation of the Ordnance Survey.

Though those responsible for 'trig. and levelling' rightly point out that theirs is the backbone without which no detail survey could be made, it is the linear, published version of the detail survey which is what 'maps' mean to most of us, and eighteenth-century map-making was far less tied to rigid triangulation than it would be in the nineteenth. Thus though the Ordnance triangulation would only extend into South-West England from 1795 onwards, mapping which has a strong claim to be called 'Ordnance Survey' had been made of Plymouth and its environs a decade earlier. Once again, the initiative seems to have come from the Duke of Richmond, who had schemes for additional fortifications around Plymouth and who hired William Gardner to make a six-inch (1:10,560) survey, extending twelve miles inland and eight miles to east and west of the town.[2] Gardner then moved eastwards to survey part of south-eastern England, and further Ordnance mapping of the South West only got under way in 1801, but the claim to precedence was established.

Mapmaking was as much a part of defence infrastructure as fortification and shipbuilding, and by the time that war with revolutionary France broke out in 1793 a start had been made with mapping south-eastern England. This was the most militarily important part of the country, and it was only in 1800 that its mapping was complete. Like Roy's map of Scotland of 1748-55 and Gardner's survey of Plymouth, of 1784-6, it remained in manuscript, an office reference tool; the emphasis was on relief and land-use, i.e. 'strong ground' and potentially helpful or obstructive ground. It was a survey with its emphasis on qualitative rather than quantitative information: this was still very much the era of not shooting till one saw the whites of one's enemy's eyes. The second most important area militarily was the South West, and between 1801 and *c.*1810 the Ordnance surveyors were wholly occupied with mapping this part of the country. Though the two-inch (1:31,680) scale had been used latterly in the South East, the survey of eastern Devon and western Somerset was made initially at the three-inch (1:21,120) scale; perhaps this was found too extravagant, as the two-inch was used for the rest of the South West and for the continuation of the survey further afield. Most of this important manuscript archive, the 'Ordnance Surveyors' Drawings', is now in the British Library, Map Library. It must added that, though the advance over commercial county mapping was undeniable, that still this early Ordnance mapping was not perfect, and that wherever possible its evidence should be cross-checked with other contemporary mapping, such as estate maps.

Initial One-inch Publication and Early Revision

By the time that the survey of the South West was completed around 1810, the nature of Ordnance mapmaking had changed profoundly. In 1801 the survey of Kent had been engraved and published, at the one-inch (1:63,360) scale, by William Faden, Geographer to the King. This seems to have inspired the Ordnance to do their own engraving and publishing, of what would later be called the 'Old Series' one-inch map, starting with four sheets of Essex, published in 1805, and continuing in 1809 with eight sheets covering most of Devon and parts of Somerset and Cornwall. The 1809 octet included a reduction of Gardner's 1784-6 Plymouth survey. Four sheets covering Dorset appeared in 1811, followed by five of Cornwall, nominally in 1813, though from 1811 to 1816 the sale of the maps to the public was forbidden by the War Office. Though the Ordnance maps were in the general line of eighteenth-century county map-making, what particularly distinguished them was the general attention to minute detail, and particularly to the depiction of relief. At this time the influence of the South West on the Ordnance Survey was more than merely geographical: its superintendant, Colonel William Mudge, and its principal hill draftsman, Robert Dawson, were both natives of Devon.

In 1820 the mapping of the South West was effectively completed with the publication of Old Series sheet 28, covering Lundy. Sheet 28 in its original form is now rarely met with, as it had to be withdrawn almost immediately and re-engraved. The Admiralty hydrographers were surveying in the neighbourhood and were perturbed to find that 'the direction of this Island as given by the Ordnance Survey is quite incorrect'.[3] The surveying vessel was H.M.S. *Hasty*, but it was the Ordnance who were proved to have been guilty of hasty work. Captain Thomas Colby, who had just taken charge of the Survey, had already discovered faulty work in the unpublished plans of eastern England, and immediately sent his deputy to resurvey Lundy. For the next fifteen years almost the whole of Ordnance Survey work in Britain lay in revising or resurveying defective unpublished work; demands from potential users meant that the priority lay in publishing new mapping rather than in revising existing mapping.[4]

However, some revision of published Old Series sheets was undertaken, including those in the South West. The spur to this came from two sources. On the one hand the Admiralty wanted a better depiction of the coast with which to connect their hydrographic surveys, and on the other there was geology, the great 'growth science' of the first half of the nineteenth century. In 1832 the Board of Ordnance was approached by Henry De la Beche with an offer to 'colour geologically' the eight sheets of Devon, for £300. De la Beche was a professional geologist in every save the financial sense, being a somewhat impecunious landed gentleman. Unofficial geological work had been going on in Cornwall and Somerset, and Devon may have appealed to De la Beche as being geologically unexplored. His geological survey covering the eight Devon sheets was completed in 1835, and the Ordnance accepted his offer to extend the work over the rest of the country. De la Beche moved into Cornwall.[5] One of the difficulties was that, partly no doubt due to the lapse of time since the two-inch field survey, and partly due to the defective nature of the work, it was difficult to make a satisfactory geological survey owing to the difference between paper and actual landscape. The problem was perhaps particularly acute around Penzance: at any rate sheet 33 of the Old Series was wholly revised, and republished in 1839. Though the manuscript Ordnance Surveyors' Drawings and the published unrevised Old Series maps should be used with some caution, particularly when accurate positioning is in question, so too should the revisions of the late 1830s. To give a well-known instance, were Ordnance Survey maps to be believed, Croyde church in north Devon moved away from the sea between the surveys of 1804-5 and the revision of *c*.1837, and then moved a whole quarter of a mile landwards by the time of the next survey in the 1880s, notwithstanding that the erosion of the cliffs to seaward of the church would tend to lessen the distance rather than increase it.[6] But at least an attempt was made at revision: by the middle of the century a considerable number of new roads had been constructed which would not appear on Ordnance Survey maps until the late 1880s.[7]

The Development of Large-scale Mapping

By the time that these limited revisions were being undertaken the Ordnance Survey in Great Britain was about to undergo a transformation as radical as that of changing from being an agency which merely produced manuscript maps to being one which published them as well. Since 1824 it had been mapping Ireland at the six-inch scale for fiscal reasons, the scale having apparently been chosen because a Commons Select Committee liked the look of the six-inch work in Kent, though if one is to trace the six-inch's family tree back the Plymouth survey of course takes seniority. Various geological interests decided that in Britain the six-inch would be superior to two-inch survey and one-inch publication and they persuaded the Board of Ordnance and the Treasury to think the same way. Unfortunately, financial stringency meant that the six-inch did not move forward quickly enough to satisfy the Scots, in particular, and in 1851 an attempt to abandon it in favour of reverting to the two-inch/one-inch set-up started the 'Battle of the Scales', the outcome of which was the adoption in 1854-6 of the standard scales of 1:500 (126.7 inches to one mile) for urban surveys, 1:2500 (25.344 inches to one mile) for rural survey of cultivated areas and six-inch for uncultivated districts. In addition, six-inch mapping was to be published derived from the 1:500 and 1:2500, and one-inch mapping was to be derived from the six-inch. The first priority was the unsurveyed part of northern Britain, but in 1863 the resurvey of the whole of southern Britain was authorised.[8]

Most of the South West was amongst the last areas to be resurveyed at 1:2500, mostly in the 1880s, but it had had a foretaste of the bright new cartographic future much earlier. As in the late eighteenth century the motive was defence. In 1848 there was considerable anxiety as to whether national defences would be adequate in the event of a French invasion and, as part of preparations for further fortification work, a large number of Ordnance surveyors were temporarily transferred from northern England to make a survey of Devonport and its environs. They recorded the measurements of the ground in their field books, but then the immediate emergency passed, and for the time being the data in the field books remained unplotted on paper.[9] Indeed, the first fruits of the Devonport survey seem to have been a ten-foot scale (1:528) plan, prepared at the request of the Local Board of Health. In the early 1840s the Ordnance Survey had imported the five-feet (1:1056) scale from Ireland with the six-inch, but the General Board of Health insisted that nothing less than the ten-foot scale would do for plans prepared to execute public sanitation schemes. The Ordnance Survey grudgingly obliged, and in 1850-53 made about thirty 1:528 surveys for public health purposes, including one of Torquay. These surveys were made notionally on a repayment basis, but unfortunately Captain Beatty, the Royal Engineer officer in charge of the public health work, does not seem to have been very good at estimating the likely costs, which usually proved to be about twice or thrice what he had estimated, and which the towns declined to pay in full. Devonport and Plymouth got off more lightly, as all that was necessary to oblige them was to plot the 1848 survey, already paid for by the national exchequer. These sanitary surveys included very minute detail, including existing sanitary features such as water taps, sewer grates and lamp-posts, health hazards such as cow-houses, industrial premises and rubbish dumps, and moral hazards such as public houses. The Ordnance Survey specimens were very highly finished, accounting in part, it was alleged, for the excessive cost.[10] The majority of the public health plans remained in manuscript (that of Torquay seems to be lost), but that of Devonport and Plymouth was engraved, no doubt because of its military applications, although a number of the sheets seem to have been withheld from sale. The 1:500 of Plymouth and Devonport was complemented by a twelve-inch (1:5280) map of the district, in eight sheets, which appears to be the only Ordnance Survey use of this scale. It is difficult to say why it was employed: surveying at twelve-inch was suggested by Lt-Colonel Lewis Hall, then head of the Ordnance Survey, as an attempt to salvage personal dignity when his opponents in the Battle of the Scales suggested 1:2500, but by the time the twelve-inch of Plymouth appeared Hall had been succeeded by Major Henry James, who was an enthusiastic advocate of the 1:2500.

It was James who presided over the initial 1:2500 work in the South West, though he was in his grave by the time that the bulk of it was under way. In 1859 there was further apprehension of French intentions, and the then Prime Minister, Lord Palmerston, embarked on an extensive fortification programme, dubbed 'Palmerston's Follies'. As well as further 1:2500 surveys around Plymouth, in the early 1860s surveys at that scale were made around Falmouth, Torbay and Weymouth and Portland (Figure 6.1). As was the practice at this time, the 1:2500 plans were published by parishes, accompanied by books of reference giving land utilisation information; the six-inch was published as far as possible, resulting in several incomplete sheets. As well as the published mapping, there was a series of detailed plans of fortifications which, though printed, remained confidential.

In the late 1860s there was considerable pressure from mining interests for the completion of the 1:2500 and six-inch resurvey and it was the priority given to mining districts which accounted for 1:2500 survey in Cornwall beginning in earnest in 1875, with the rest of the South West being surveyed between 1882 and 1889. Devon had the unenviable distinction of being the last county to be resurveyed for the 1:2500 (or six-inch in uncultivated areas), being completed only in 1888-9. Towns over 4000 population were mapped at the 1:500 scale: Exeter was mapped at this scale twice, in 1875-6 as a repayment job, and again in 1888 as part of the county survey.[11] The 1:500 surveys in the South West are listed in Appendix 1. For some reason the military two-inch surveys of the 1800s had left out the Isles of Scilly, which were surveyed by the Ordnance Survey for the first time in 1887. Between 1887 and 1894 the 'New Series', derived from the 1:2500/6-inch resurvey, replaced the 'Old Series' one-inch map. The one-inch was on national sheetlines, unlike the larger-scale surveys, which were on three different county systems in the South West; one system sufficed for Dorset and Somerset (though sheets along their mutual border bore two different numbers), but Cornwall and Devon each had its separate system. Since 1872 1:2500 sheets had been published by counties rather than by parishes, and from 1882 the six-inch was published in quarter sheets directly photo-reduced from the 1:2500 (Figures 6.2 and 6.3). One-inch and larger scales alike used the Cassini projection.

Small-scale Opulence and Large-scale Squalor

By the time that the resurvey of the South West was complete the earlier, defence-inspired, large-scale mapping was out of date, and in 1892 the 1:500 and 1:2500

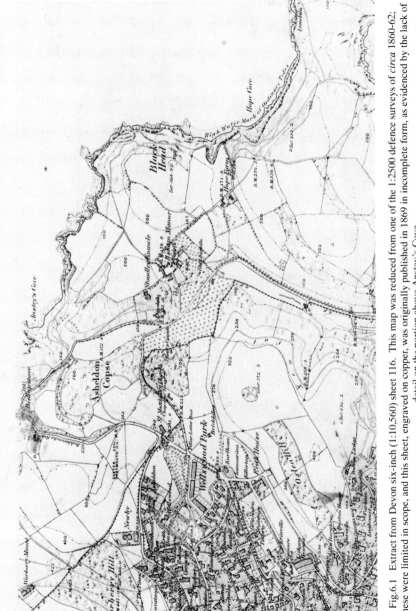

Fig.6.1 Extract from Devon six-inch (1:10,560) sheet 116. This map was reduced from one of the 1:2500 defence surveys of *circa* 1860-62; these were limited in scope, and this sheet, engraved on copper, was originally published in 1869 in incomplete form, as evidenced by the lack of detail on the portion above Anstey's Cove.

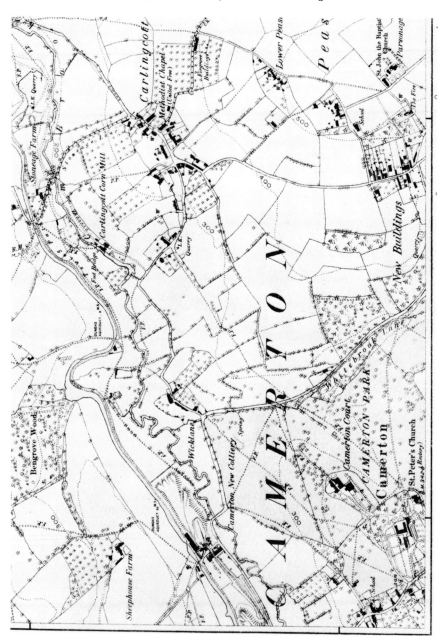

Fig.6.2   Extract from Somerset six-inch (1:10,560) sheet 20 N.E., surveyed 1884. This sheet was surveyed at the 1:2500 scale in 1884 and produced by direct photographic reduction from that scale as a mass production economy measure.

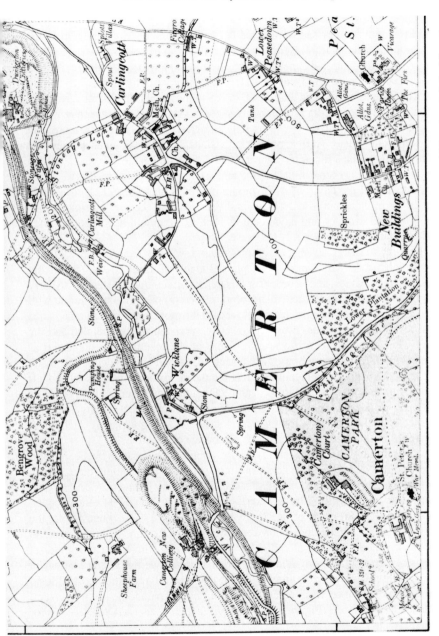

Fig.6.3 Extract from Somerset six-inch (1:10,560) sheet 20 N.E., revised 1929. This covers the same area as 6.2 but, though derived from the 1:2500, was drawn independently; hence the bolder style. By the time that this sheet was revised, financial pressure was causing the Ordnance Survey to restrict rural revision to urban hinterlands and semi-industrial areas such as this. Because of the coal deposits, this was one of the first areas in Somerset to be resurveyed for the National Grid 1:2500 in the late 1950s.

around Plymouth was revised, some time before a general national revision at this scale was authorised in 1894. Henceforth 1:2500 was to be the largest scale which would be revised at national expense; towns wishing to have their 1:500 revised at their own expense could do so, but only a few did, and none of these was in the South West. The scheme of revision authorised in 1894 envisaged two separate groups of revision, one for the one-inch, which envisaged no sheet being more than fifteen years out of date, and one for the larger scales, so that they would not be more than twenty years out of date. At first, one-inch revisions took place at much shorter than fifteen-year intervals: a first took place in 1894-7, a second in 1903-9, and a third in 1912-14. The first revision of the 1:2500 and six-inch took place in 1900-7, and a further revision of the environs of Plymouth only was made in 1912, presumably primarily for military purposes. (Incidentally, because of security considerations, mapping around military and naval centres such as Plymouth and Portland often lacked contours and altitudes.)

By that time the large-scale revision programme was in some disarray. One of the provisions of Lloyd George's 'People's Budget' of 1909-10 was for taxing increases in land values, and in order for this to take effect it was necessary to carry out a systematic valuation by tenements of all landed property, in conjunction with Ordnance Survey 1:2500 maps. In order to record the more important changes since the last revision, the regular 'twenty-year' cyclic revision was suspended and the revisers were switched to urban areas. Some of the results were published as 'Special Editions', with new houses left unshaded, certain broken detail not 'made good', and other signs of an incomplete job, but the majority seem to have been left in manuscript and handed over to the Land Valuation Department, which made its own copies. The Land Valuation Department also needed 1:1250 enlargements of some sheets, as the 1:2500 was too small a scale to show adequately such small landholdings as individual terrace houses, for instance (Figure 6.4).[12]

No sooner had regular large-scale revision resumed its course than the First World War broke out. Naturally, being a part-military, part-civil department, there was an immediate exodus of manpower, and this was followed by stringent restrictions on Ordnance Survey spending. The practical result of this was that by the time a second revision of the 1:2500 plans of the South West began in 1923, around Poole, the 'twenty-year cycle' had practically broken down, and by the time that the last urban revisions were made in Cornwall in 1933 it was a dead letter; in Dorset and Somerset the plans of every small town, and some larger villages, were revised, whereas in Devon smaller towns were left unrevised, and in Cornwall even an important local centre such as Penzance was denied a revision, though paradoxically a few rural hinterlands were mapped (Figure 6.5). The revised 1:2500 also served as a basis for a revised six-inch.

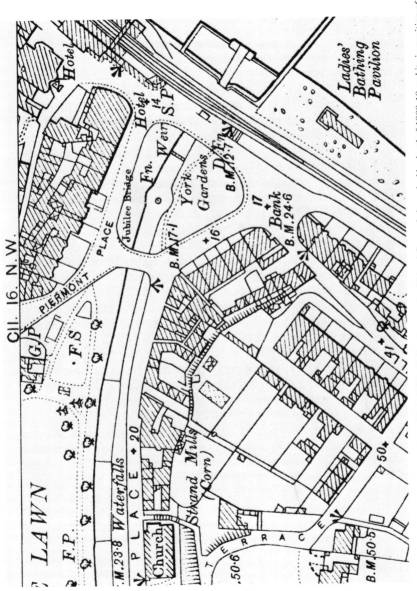

Fig.6.4 Extract from Devon 1:1250 sheet 102.16.S.W., 1912. As a result of Lloyd George's 'peoples' budget' of 1909-10 and its imposition of a tax on the unearned increment of land values, a massive land valuation exercise was carried out during the next five years, which included the preparation of annotated Ordnance Survey maps showing land ownership. The largest up-to-date Ordnance Survey coverage of urban areas was 1:2500 which was photo-enlarged to 1:1250 to enable land ownership to be mapped.

Things were somewhat better with the one-inch. Colour-printed one-inch maps of the South West had started to appear in 1899, and in 1914 the intention was that the third revision of the New Series map would appear in an elaborate colour scheme needing about twelve printings, with an eleven-fold road classification and relief shown by contours, hachures and layer tints. The war put an immediate stop to this, and only one proof sheet is known to survive, covering Torquay and district.[13] Instead, there appeared in 1918-19 the 'Popular Edition', which retained the elaborate road classification and contours, but jettisoned the hachures and layers. The first sheets printed covered Exmoor, Bridgwater and Tiverton districts. The 'Popular' seems to have been so called because it was envisaged as a down-market alternative to the de-luxe style proposed in 1914, but, apart from a few 'Tourist' sheets on irregular sheetlines, no de-luxe alternative to the 'Popular' appeared. It was certainly 'Popular' with the public, no doubt as much because of the growing numbers of motorists and walkers as for its cartographic merits. In the 1930s it provided the base-map for the first published maps of the Land Utilisation Survey.

The main difficulty with the 'Popular' was that the public liked it more than did its makers. It was produced by making transfers from engraved copper plates onto stone: the copper plates were wearing out, engraving was a nearly defunct art, and the image on lithographic stones was apt to lose sharpness. The Ordnance Survey had used photo-reduction since 1855, and photo-zincography (an early form of photo-lithography) for the six-inch since 1881 and for the 1:2500 since 1889, and started to use it in the 1920s for the one-inch 'Popular' of Scotland. As the Scottish and English 'Populars' were intended to harmonise, the Scottish map was in a quasi-copperplate style, which was not very well suited to photo-lithographic map production. The result was a redesigned map with specially designed lettering, and as the revisers worked, roughly, from Lands End to John O'Groats at this time it was natural that the first sheets of the new 'Fifth (RELIEF) Edition' should be of the South West. It was decided to revive the elaborate colouring proposed in 1914, but improvements in printing technology enabled it to be obtained with a greatly reduced number of printings. The first sheet of the new map appeared in 1931 and covered Plymouth and district. It had been the subject of a number of colouring experiments during the previous couple of years. The critics liked it, but the public did not, perhaps partly because it cost about 20 per cent more than the 'Popular', and partly because of tinkering with the sheetlines. Publication was fitful, and when a 'non-relief' version was offered in 1935 the public went for that. Further production of the 'Relief Edition' was abandoned in 1936, before it could reach much of Somerset or any of Dorset, which were covered only in the 'non-relief' style, and with a further re-arrangement of sheetlines. The Fifth Edition one-inch differed from its predecessors in being cast on a Transverse Mercator projection and in carrying a co-ordinate reference system, or 'grid', though not a very 'user-friendly' one. Sheets 144 and

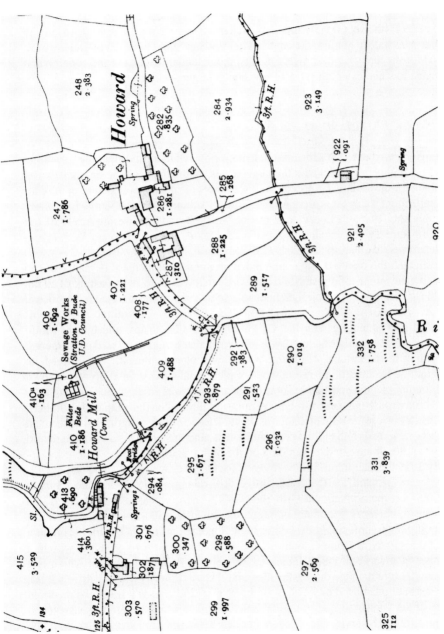

Fig.6.5   Extract from Cornwall 1:2500 sheet 5.3. revised 1932. This is a good example of the earlier twentieth-century 1:2500 style and of rural 1:2500 topographic content at all periods.

145 of the Fifth (Relief) were supplemented by gazetteers, a scheme dropped in 1935 as the sales hardly ran into dozens.[14] The 'Popular' and the Fifth continue to have value for the historian in showing at least the essential features of the country, and mitigate to some extent the lack of revision at larger scales.

Viscount Davidson to the Rescue

Though by the later 1930s the one-inch map appeared to be in something of a mess, it was still in better health than the large-scales. More than ever before there was a demand for up-to-date, large-scale maps, to meet the demands of planning in order to implement such legislation as the Town Planning Act 1925, the Electricity (Supply) Act 1926, the Local Government Act 1929, and the Housing Act 1930, but the Ordnance Survey response was perforce meagre. It was an intolerable situation. A departmental committee under J.C.C. Davidson, M.P. (later Viscount Davidson) investigated the matter and recommended an immediate increase in manpower to clear off the arrears of revision and then, once that was done, the recasting of all the Ordnance Survey's maps on sheetlines related to a national grid based on the metre, on the Transverse Mercator projection. The Davidson Committee also recommended experimenting with the 1:1250 and 1:25,000 scales, and that in future 'continuous revision' be adopted, whereby each plan would be revised on its merits, according to the amount of change which took place within it. The immediate programme resulted in a large number of piecemeal surveys in 1936-40 for town and country planning purposes; some of the mapping was urban, including Penzance and its environs, and some rural, for example to the east of Barnstaple. In addition, the institution of 'Air Raid Precautions' in 1938 led to a hasty updating of the six-inch to show new roads and buildings around towns, and this provides a valuable record of the approximate extent of building on the outbreak of war.

By 1939 the clearing-off of the revision backlog was well advanced and preparations were in hand for the considerable task of recasting all Ordnance Survey mapping on metric National Grid sheetlines. The outbreak of war delayed this somewhat, but work continued on preparing a metric-gridded version of the Fifth Edition, rather confusingly called the New Popular, and in 1943 work began on a 1:1250 resurvey of Bournemouth. By 1947 1:1250 resurvey was under way in a number of towns and cities, including Bristol and Plymouth, both of which had suffered severe damage during the war, and it started shortly afterwards in Exeter, another badly damaged city, and in Bath. The 1:1250 resurvey was to be the main work of the Ordnance Survey until about 1960, after which the emphasis switched to rural 1:2500 resurvey and revision, but before the full rural programme could be put into effect it was necessary to try out various revision and resurvey techniques.

It was also necessary to investigate what 'continuous revision' would amount to in practice. To this end, several experimental 1:2500 areas were started, including Essex, as averagely rural, and Devon as very rural. In Essex a method of resurvey using aerial photographs was tried (the 'Chelmsford' method), whereas in Devon 'overhaul' was tried, which in effect was a revision of the old County Series plans, suitably adjusted to new geodetic standards to eliminate errors which had crept in over the years, and recast on the National Grid (the 'Cotswold' method). Once 'overhauled' and revised, the plans were redrawn and republished. In fact, as survey was by blocks on National Grid sheetlines, the Devon block excluded that part of the county roughly to the north-east of a line from Exeter through Tiverton and Barnstaple and Ilfracombe, but included part of eastern Cornwall. Why Devon was chosen is not at present clear, but it may have been a matter of convenience, due to its proximity to Plymouth and Exeter, and including, in Dartmoor, a tract of basic six-inch survey.

Work got under way in 1948-9, but was not completed, even within the initial block, until 1960, and though by 1955 survey was under way in all the south-western counties it was not until 1980 that all the County Series plans were replaced by those on the National Grid. Because of obligations under the Act nationalising the coal industry in 1947, much of the 1:2500 work in the late 1950s was undertaken in the coalfields, including the Somerset coalfield, around Radstock. Appendix 6.2 gives approximate 1:2500 National Grid resurvey or revision dates for south- western towns. By that time there had been a number of developments. By 1952 it was becoming apparent that the detail within many 1:2500 plans had changed very little since the previous revision in the 1900s, and it was decided to adopt a technique known as 'by-passing', whereby each sheet would be reconnoitred, and, were the amount of change small - less than the equivalent of ten new houses - the plan would be 'by-passed', and the old County Series mapping would be republished on National Grid sheetlines. The whole concept of the by-passed plan struck at the original intention of the Davidson proposals, whereby the mapping of the whole country would be brought up to to a uniform modern standard. It is to be suspected that at least some of the staff involved in producing the by-passed plans were not very happy about the dilution of standards, and in any case the by-passed plans created problems of their own, notably the assimilation of parcel numbers and acreages where by-passed plans adjoined properly overhauled plans. Further pro- duction of by-passed plans was abandoned in 1956. They were published in a number of counties, but by far the greatest proportion was in Devon and eastern Cornwall, sometimes amounting to over 50 sq km in a 100 sq km block. Some idea where they occur can be gathered from those 1:25,000 Second Series sheets which include a 1900s survey date. The by-passed plans are a nuisance both to the local historian and to the contemporary user, for the threshold for revision of rural 1:2500 plans has been raised to such a level that many of the by-passed plans have failed to come anywhere near any sort of 'Continuous Revision'.[15]

The 1:1250 urban plans have also had the threshold for new editions under the 'Continuous Revision' procedure raised progressively, so that in practice new editions are not produced so frequently today as they were in the 1950s. For 1:1250 plans it is 300 'house-units', for 1 x 1 km 1:2500 plans 500 units, for 2 x 1 km 1:2500 plans 450 units, for 1:10,000 basic scale plans 500 units, and for publication of a SIM about 50 units.[16] (SIM stands for 'Survey Information on Microfilm'; a partially revised version of the sheet is issued on a microfilm aperture card, enabling printouts to be made). What this means in practice is that the post-war redevelopment of Exeter and Plymouth is well covered. A few examples may be given. Sheet SX 4754 NE, which includes most of Armada Way in Plymouth, was surveyed in January 1951 and revised six times between November 1952 and June 1968; since then two SIMs have been published, the latest being in November 1988. Sheet SX 9292 NW, which includes Exeter cathedral and the redeveloped part of the High Street, was surveyed in April 1950, and revised five times between December 1951 and February 1976; since then there have been two SIMs, the latest being dated November 1988. Sheet SS 4526 NW, part of Bideford, was resurveyed in February 1979; so far no SIM has been published. Although frequent new editions may have been within the spirit of 'Continuous Revision', in practice there were problems, not least because of the lead-time between revising in the ground and publishing the revised plan: if a plan was revised at 18-month intervals, it might take 9 months to go through the production line, so that it was only really up-to-date for 9 months before a further revision was undertaken. Thus during the 1950s and 1960s there was a gradual lengthening in the interval between new revisions and editions. A further complication is the amount of work which justifies sending a surveyor to the ground. More recently, the problem of reconciling 'Continuous Revision' with publication has been got over by the issue of SIM. From the historian's point of view the SIM versions are not ideal, as the only date borne is that of publication, not of fieldwork, but from the contemporary user's point of view they have the advantage that they reduce the discrepancy between the mapped and the actual landscape.

Experience has shown that though 'Continuous Revision' works well in urban and other areas of comparatively rapid change, it is less satisfactory in rural areas, where the amount of change is rarely sufficient to justify the publication of new plans. Ordnance Survey policy at present is that all plans are to be revised at least once every forty years and, as an alternative to 'Continuous Revision', in recent years 'sweeping' in blocks of about 200 sq km has been tried, mostly in areas such as the South West which were amongst the first to be covered by National Grid 1:2500 mapping.[17] Small towns mapped at 1:2500 have usually had two to four new editions published of their 1:2500 plans, and some have been resurveyed at 1:1250 scale. To give a single example, Bideford was revised at 1:2500 in 1956-7, 1966 and 1973, before being resurveyed at 1:1250 in 1979.

Derived Maps, A ir Photo Mosaics and Provisional Editions

The 1:1250 was photographically reduced to provide continuous coverage at 1:2500 of both urban and cultivated rural areas (though early (c.1948-52) derived 1:2500 plans were redrawn), and the 1:2500 was used to provide the raw materials for new six-inch and 1:25,000 mapping. The by-passed plans apart, the National Grid 1:1250 and 1:2500 were better than anything seen before (the nineteenth-century predecessors perhaps scored on minor details, but not on general execution), but they took a long time to prepare. As a result various makeshifts were adopted at various scales. In 1946-7 the RAF flew a good deal of aerial photography which could be used by the Ordnance Survey, and the idea was conceived of the 'Air Photo Mosaics', which were effectively groups of air photographs published on National Grid sheetlines. Ninety of a projected 103 air photo mosaics were prepared for Exeter, and twenty-three of a projected 112 for Plymouth. Six-inch mosaics were also produced for some rural areas, though the only ones to be produced in the South West were for south-east Dorset and around the mouth of the Exe. These are listed in Appendix 6.3. Though lacking the clarity and definition of a conventional map, they are valuable as guides to land utilisation in the 1940s, and the date of the photography is given at the foot of each mosaic. There is also the convenience of a small overlap between mosaics. For historians, there is the drawback that they had a very limited circulation and are unlikely to be found outside national or copyright libraries; probably more were given away for copyright purposes than were actually sold. Though it might be thought that aerial photography would long since have replaced conventional mapping, in practice there are various difficulties. One is obscuring by clouds; another is relative unfamiliarity with, and demand for, the product; a third is that whereas undesirable details on a conventional map are simply not surveyed and drawn, on an aerial photograph they have to be painted out. The Air Photo Mosaics were produced at the time of the Cold War, and changing rules for 'security deletions' meant that the entire stock had to be withdrawn from sale in 1954, and the small demand did not justify the necessary 'security check'.[18]

Much more widely disseminated than the air photo mosaics were the various 'provisional editions' of the six-inch and 1:25,000 maps. In 1943 work started on publishing a civil edition of the 'ARP' revision of the six-inch of 1938-9, retaining county sheetlines, but adding the National Grid, and many of the earliest sheets were of the South West. Very conveniently for the historian, the buildings added in 1938-9 were shown without the usual hatching. This was followed by cartographically new mapping at 1:25,000 (2.53 inches to one mile), of unprecedented clarity, published in 10 x 10 km sheets. Experiments at this scale had been recommended by the Davidson Committee, but a direct reduction from the six-inch produced for the military in 1940 proved to be so useful to planners and others that it was decided to press ahead regardless with a completely redrawn map at this scale. Unfortunately, like the parent 'Provisional Edition' six-inch sheets, it was a mixture of limited revision of the 1930s added to a fundamentally 1900s revision, and though it had its uses for planners and educationalists, the public liked it rather less at first. As a record of the landscape the 'Provisional Edition' or 'First Series' 1:25,000 is of very limited importance, but it can be useful in other ways; for example, it shows field boundaries before the extensive removal of hedges after 1945, and thus can be useful for plotting information from enclosure and tithe maps. As its title indicates, it was intended as something to be going on with until a 'Regular Edition', based on the new National Grid large-scale mapping, could be published.

In 1953-61 the one-inch Seventh Series replaced the New Popular in south-west England, sheet 190, published in 1961, being the official 'final sheet'. The revision for the Seventh Series was also used for partial revision of the 1:25,000 and the six-inch. Publication of a redrawn six-inch, based on the 1:1250 and 1:2500 National Grid mapping, had begun around Plymouth in 1954; in the late 1950s and early 1960s a 'Provisional Edition' of the six-inch on National Grid sheetlines covered those areas awaiting the provision of National Grid mapping at larger scales. From the historian's point of view the 'Provisional' six-inch has its uses for indicating in more detail those landscape changes (e.g. new building and afforestation) which would appear on the one-inch, but such features as field boundaries were left unrevised.

In the mid-1950s it was envisaged that the new six-inch and 1:25,000 'Regular Editions' would be produced entirely separately, and a block of eleven redrawn 1:25,000 'Regular Edition' sheets around Plymouth were published in 1956. The mapping improved on the 'Provisional Edition' by adding green for woods. In the event the eleven sheets around Plymouth remained the only ones of their sort, as it was decided to experiment with drawing the six-inch more boldly in order that its linework could be used for the 1:25,000 'Second Series', as the 'Regular Edition' was renamed. It was also decided to experiment with much larger sheets, on irregular sheetlines, and a 20 x 15 km sheet, '856', covering Ilfracombe and Lundy, based on First Series material, appeared in 1960. 'Sheet 856' remained the only one of its sort to reach the public, as it was decided subsequently to publish the Second Series in 20 x 10 km sheets. Though a considerable number of the early Second Series sheets were of Devon and eastern Cornwall, the very last to be published, 1362, in November 1989, replaced the last of the eleven 'Regular' sheets of 1956

As a general map of the countryside the 1:25,000 Second Series is invaluable, being as it is the smallest scale to show field boundaries, but as a record of the landscape it is subject to somewhat similar limitations to those which affect the First Series, in that it is based on surveys made and revised at various times. In the South West there is the particular hazard that it includes material taken from the 'by-passed' 1:2500 plans. The six-inch and 1:10,000, particularly editions published before about 1974, are more satisfactory in that respect, but when dating is crucial there is no alternative to refer to every available edition of the basic scale, 1:2500 or 1:1250.

Conclusion

The South West, in common with regions of comparable extent elsewhere in Britain, has its own distinctive pattern of Ordnance Survey coverage over the past two centuries. In some ways it has benefited from having been the location of some of the earliest Ordnance Survey work, at both small and large scales, though the comparatively late date at which large-scale coverage was completed meant that it was at a disadvantage when its turn for revision was due between the two world wars. To some extent this was compensated for by the experimental 1:2500 work in the 1950s. In the twentieth-century Ordnance Survey mapping of the South West is perhaps of greater national than regional significance and interest; the region has had more than its share of experimental maps and map series which were abandoned incomplete, and this pattern has continued in recent years.

Notes and References

1. The next four paragraphs are mostly based on the introductory essays by J.B. Harley and Y. O'Donoghue in H. Margary, *The Old Series Ordnance Survey*, vols I and II (Lympne Castle, 1975, 1977).
2. Copies in Public Record Office, London, MR 1199.
3. P.R.O. OS 3/260, p.112.
4. J.B. Harley, 'Error and Revision in Early Ordnance Survey Maps', *Cartographic Journal*, 5 (1968), 115-24.
5. J.B. Harley, 'The Ordnance Survey and the Origins of Official Geological Mapping in Devon', in K.J. Gregory and W. Ravenhill (eds.), *Exeter Essays in Geography* (Exeter, 1971), 105-23.
6. A.P. Carr, 'Cartographic Record and Historical Accuracy', *Geography*, 47 (1962), 135-44.

7. J. Bentley, 'Research into Roads in Somerset, using the Ordnance Survey', *Sheetlines* (forthcoming).

8. For mid-nineteenth-century developments see Richard Oliver, 'The Ordnance Survey in Great Britain 1835-1870' (unpublished D.Phil. thesis, University of Sussex, 1986).

9. See P.R.O. WO 47/2165, pp.15861-2; WO 47/2193, pp.11420-2; WO 47/2219, pp.4743-5; WO 47/2225, pp.6940-2.

10. J.B. Harley, 'Cartography and Politics in Nineteenth-Century England: the case of the Ordnance Survey and the General Board of Health, 1848-1856', unpublished paper presented at the VIIIth International Conference on the History of Cartography, Berlin, September 1979.

11. J.B. Harley and J.B. Manterfield, 'The Ordnance Survey 1:500 Plans of Exeter, 1874-1877', *Devon and Cornwall Notes and Queries*, XXIV (1978), 63-75.

12. See Public Record Office classes IR 125 and IR 128.

13. This map is now in the 'Ordnance Survey Specimen Drawer' in the map room of the Royal Geographical Society, London.

14. P.R.O. OS 1/49.

15. P.R.O. OS 1/496.

16. Ordnance Survey to writer, 5 September 1990.

17. Ordnance Survey annual reports 1982-3, 7-8, 1983-4, 3.

18. P.R.O. OS 1/304, OS 1/305, OS 1/306.

APPENDIX 6.1

*Ordnance Survey urban surveys at 1:1250, 1:528 and 1:500
(excluding direct enlargements from 1:2500 to 1:1250)*

Columns:
1, town, or group of towns;
2, scale;
3, number of plans, [omitted for some National Grid surveys, particularly in conurbations; many 1:500 revisions were partial];
4, (county series only) MS = manuscript, E = engraved, C = lithographed or zincographed with hand-coloured buildings, L = lithographed or zincographed with buildings hatched;
5, survey/revision date;
6, type, meridian and sheet-numbering system (1:528/500 only): LS = local meridian and sheetlines system, made for local board of health, usually with additional sanitary information (all 1:528 were on local meridians and sheetlines, all but a very few early 1:500 were on county sheetlines, numbered as subdivisions of the 1:2500, and all post-1943 surveys are National Grid); (National Grid series only) R = resurvey of town originally revised for NG at 1:2500;
7, veracity of information: RD = information from progress diagrams in OS annual report; V = details verified from BL copies, VR followed by date = bracketed date is given on the maps, but main date is thought more authentic, VRX = no survey date given on maps.
Compilation: unless noted otherwise in column 7, details of county series surveys are from the OS *Catalogue* of 1897, and H.S.L. Winterbotham, *The Replotted Counties* (London, HMSO, 1934), and details of National Grid surveys are from OS *Annual Reports*, monthly publication lists and the Plan Availability Index ('Plandex').

1	2	3	4	5	6	7
Barnstaple	1:500	25	L	1885		VR (88)
	1:1250	41		1976-8		
Bath	1:500	86	L	1882-3		
	1:1250	79		1949-53		*

* 55 of 99 projected Air Photo Mosaics prepared.

1	2	3	4	5	6	7
Bideford	1:500	10	L	1885		VR (86)
...Westward Ho & Northam	1:1250	50		1977-80	R	V (78-9)
Bodmin	1:500	18	L	1880		
Bridport	1:500	16	L	1886-7		
Bridgwater	1:500	33	L	1885-6		
	1:1250	45		1965-7		
Bristol	1:500	182	L	1879-82		*
	1:1250	445		1946-53		+ $

* 1:528 plan tested by OS in early 1850s.
+ Bristol was resurveyed early on because of extensive bomb damage and as survey control was already in place. See P.R.O. OS 1/170.
\$ Air Photo Mosaic coverage proposed but not prepared.

1	2	3	4	5	6	7
Brixham	1:500	5	C	1862		
	1:1250	28		1980-2	R	
Camborne	1:500	9	L	1876-7		
... & Redruth	1:1250	53		1962-7		
Clevedon	1:500	28	L	1882-3		
	1:1250			1988?	R	
Crediton	1:500	13	L	1886		
Crewkerne	1:500	20	L	1885		
Dartmouth	1:500	14	L	1885		
Dawlish	1:500	9	L	1887		VR (87-8)
Exeter	1:500	41	L	1875-6		*
	1:500	41	L	1888		*
	1:1250	93		1948-53		+

* See J.B. Harley and J.B. Manterfield, 'The Ordnance Survey 1:500 Plans of Exeter, 1874-1877', *Devon and Cornwall Notes and Queries*, XXIV (1978), 63-75.
+ 90 of projected 103 Air Photo Mosaic prepared.

1	2	3	4	5	6	7
Exmouth	1:500	17	L	1887-8		
	1:1250	30		1950-2		
Falmouth	1:500	10	L	1887-8		
... & Penryn	1:1250	46		1967-9		
Frome	1:500	27	L	1884		
	1:1250	35		1983	R	
Ilfracombe	1:500	16	L	1886-7		VR (88)
Keynsham	1:1250			1986-7		

1	2	3	4	5	6	7
Midsomer Norton						
& Radstock	1:1250	48		1980-1	R	
Nailsea	1:1250	20		1984	R	
Penzance	1:500	16	L	1875-6		
	1:1250			1987?		
Plymouth	1:500	109	E/L	1855-7	LS	VR(56-7)*
	1:500	79	L	1892-3		
	1:1250	147		1946-52		+

* The early history of the Plymouth survey is not straightforward. It appears that Devonport was surveyed, but apparently not plotted, in 1848; presumably Plymouth proper followed some years later. These surveys were primarily for military use, and only 28 sheets appear to have been for sale to the public.

+ 112 Air Photo Mosaics projected; 23 prepared.

Poole	1:500	14	L	1886		VR (87)
	1:1250	128		1950-3		
Redruth	1:500	8	L	1877		
(in Camborne survey)	1:1250			1962-7		
St Austell	1:500	7	L	1878-9		
	1:1250	38		1966-7		
St Ives (Cornwall)	1:500	9	L	1876		
Shepton Mallet	1:500	19	L	1884		
Sherborne (Dorset)	1:500	13	L	1886		
Taunton	1:500	48	L	1886		
	1:1250	57		1957-60		
Tavistock	1:500	14	L	1882-3		
Teignmouth	1:500	15	L	1887		
	1:1250			1988	R	V
Tormoham (Torquay)	1:528	?	MS	1851?		*
Torquay	1:500	18	C	1860-1		
...& Paignton	1:1250	124		1950-3		

* Made in 1851 for the Local Board of Health; present whereabouts uncertain.

Totnes	1:500	9	L	1886		
Truro	1:500	8	L	1877-8		
	1:1250	26		1987	R	
Wells	1:500	12	L	1885		
Weston-super-Mare	1:500	34	L	1883-4		
	1:1250	68		1950-2		
Weymouth	1:500	13	C	1863		
	1:1250	81		1956-8		
Yeovil	1:500	26	L	1885		
	1:1250	51		1962-4		

APPENDIX 6.2

Survey/initial revision dates for National Grid mapping at 1:2500 of small towns

Suffixes: * = subsequently resurveyed at 1:1250; V = Verified from sampling of published maps: V6 = date taken from derived 1:10,000/1:10,560 mapping.

Bodmin 1963	Minehead 1967
Brixham 1954-5	Nailsea 1971-2 *
Buckfastleigh 1954 V	Newton Abbot 1955-6 V *
Bude 1954 V	Okehampton 1954 V
Budleigh Salterton 1955 V6	Ottery St Mary 1958
Chard 1961	Padstow 1972 V
Clevedon 1964 V	Penzance 1961 *
Crediton 1964 V	Portland, Isle of, 1958?
Crewkerne 1977	St Ives (Cornwall) 1962
Dartmouth 1953 V6	Shepton Mallet 1969
Dawlish 1954 V	Sherborne 1963-4
Dorchester 1957	South Molton 1968 V
Frome 1967 *	Stratton (Cornwall) 1954 V
Gillingham (Dorset) 1964	Street (Somerset) 1967
Glastonbury 1967	Swanage 1954 V
Great Torrington 1954 V	Tavistock 1952-3 V
Holsworthy 1953 V	Teignmouth 1954-5 V
Honiton 1958	Truro 1966-7 *
Ilfracombe 1961	Wadebridge 1972
Kingsbridge 1954 V6	Wareham 1952 V
Liskeard 1977	Watchet 1971
Looe 1970	Wellington (Somerset) 1962
Lostwithiel 1970-1	Wells (Somerset) 1963
Lyme Regis 1957	Wimborne Minster 1953 V *
Lynton & Lynmouth 1974	Wincanton 1976
Midsomer Norton/Radstock 1957-9 *	

APPENDIX 6.3

Published Six-Inch Air Photo Mosaic Sheets in South-West England

Sheet numbers are of National Grid sheets.   ALL = all quarters published.

ST 80 (ALL); ST 81 (ALL); ST 82 (ALL); ST 90 (ALL); ST 91 (ALL); SU 00 (ALL); SU 01 (ALL); SX 97 NE; SX 98 NE, SE; SY 08 NE, SW, SE; SY 67 NW, NE; SY 68 (ALL); SY 78 NW, SW, SE; SY88 (ALL); SY89 (ALL); SY97 NW, NE; SY98 (ALL); SY99 (ALL); SZ 07 NW, SZ 08 NW, SW; SZ 09 NW, SW, SE.

Guide to Further Reading

The following short guide to further reading is intended for those who use old maps primarily as sources of historical data but who also wish to know more about the history of maps and mapping. The classification follows broadly the chapter organisation in this book.

1. Maps and History: concepts and approaches

M.J. Blakemore and J.B. Harley,'Concepts in the History of Cartography: a review and perspective', *Cartographica*, 17 (1980), 1-120 (*Cartographica* Monograph no. 26). The authors consider that an approach which views early maps as a language for communicating spatial ideas offers 'a stimulating and elegant structure for the broad development of the history of cartography' (preface).

J.B Harley, 'The Map and the Development of the History of Cartography' in J.B. Harley and D. Woodward (eds.), *The History of Cartography* Vol 1 *Cartography in Prehistoric, Ancient, and Medieval Europe and the Mediterranean* (Chicago, University of Chicago Press, 1987), 1-42. In this introductory essay to a multi-volume work on the history of world cartography, Professor Harley examines the historical importance of the map and reviews changing approaches to, and influences on, map history since the Renaissance and provides an extensive set of notes, references and a bibliography on these topics.

2. Maps in the Study of Local and Regional History

There are now a number of guides to historical sources and the writing of local history (see a current issue of the periodical *The Local Historian*) all of which say something about maps and their interpretation and use in the study of local and regional history. The following books deal specifically with map sources only and are oriented to the needs of local historians.

J.B.Harley, *Maps for the Local Historian: a guide to the British sources* (London, National Council of Social Service for The Standing Conference for Local History, 1972). This book contains chapters on town plans, estate, enclosure and tithe maps, communications maps, marine charts and county maps first published in the *Local Historian* between 1967 and 1969. Much more is now known about each of these categories of maps than when Professor Harley was writing in the 1960s but this has not diminished the importance of this book for historians today, particularly where matters of reading, interpreting and questioning the surface evidence of maps is concerned. Not only its title but its general approach are echoed in the two most recent guides to 'maps for the local historian': Paul Hindle's *Maps for Local History* (London, Batsford, 1988) and David Smith's *Maps and Plans for the Local Historian and Collector* (London, Batsford, 1988) appeared in the same year from the same publisher and in identical format. Indeed they cover very similar ground but with Paul Hindle stressing the practical use of maps as sources, while David Smith orientates his text a little more towards the map collector who views maps as artefacts. Both are very fully illustrated.

3. Estate and Cadastral (Property) Maps

There are chapters on estate maps in each of Harley (1972), Hindle (1988) and Smith (1988).

P.D.A. Harvey, *The History of Topographical Maps: symbols, pictures and surveys* (London, Thames and Hudson, 1980) examines the early history of local mapping on a world canvas and is particularly important for its discussion of the re-emergence of local mapping in Renaissance Europe.

Roger J. P. Kain and Elizabeth Baigent, *The Cadastral Map in the Service of the State: a history of property mapping* (Chicago, University of Chicago Press, forthcoming 1992) reviews the ways in which state agencies used maps as both practical instruments to implement policies of land taxation, re-allocation and settlement and also as graphic symbols of the power that governments came to exert over landed territory as the medieval world of feudal estates gave way to capitalist nation states in the Renaissance.

4. Enclosure Maps

In addition to the papers by Dr John Chapman cited in the references to Chapter 4, the following two books by Dr Michael Turner should be noted:

Michael Turner, *English Parliamentary Enclosure: its historical geography and economic history* (Folkestone, Dawson, 1980) is a full discussion of the process of enclosure, its spatial setting and temporal sequence.

Michael Turner, *Enclosures in Britain 1750-1830* (London, Macmillan, 1984) is an introduction to the topic prepared in association with The Economic History Society.

5. Tithe Maps

Eric J. Evans, *The Contentious Tithe: the tithe problem in English agriculture 1750-1850* (London, Routledge & Kegan Paul, 1976) is a penetrating review of the nature of the 'tithe problem' and of the mounting pressure for reform.

Eric J. Evans, *Tithes and the Tithe Commutation Act 1836* (London, Bedford Square Press for the Standing Conference for Local History, 1978) analyses the 1836 Act, discusses its implementation and reviews the source material which it generated.

Roger J.P. Kain and Hugh C. Prince, *The Tithe Surveys of England and Wales* (Cambridge, Cambridge University Press, 1985) is a general handbook to the tithe surveys and their use as sources of information on field systems, place names, land use, farming, and land ownership and occupation.

Roger J.P. Kain, *An Atlas and Index of the Tithe Files of Mid-Nineteenth-Century England and Wales* (Cambridge, Cambridge University Press, 1986) presents a county-by-county reconstruction of the mid-nineteenth-century agrarian economy and gives detailed guidance to the contents of the 'tithe file' (see chapter 5 above) of each English and Welsh parish and township.

6. *Ordnance Survey Maps*

Much of the most recent material on early Ordnance Survey maps has been published as articles or chapters in books; the most relevant of these are listed in the notes and references to Chapter 6.

J.B. Harley, *The Historian's Guide to Ordnance Survey Maps* (London, National Council of Social Service for The Standing Conference for Local History, 1964). Most of the chapters of this book were previously published as articles in *The Amateur Historian* some thirty years ago now; inevitably some errors of detail have been revealed by subsequent scholarship but the general advice about the evidence of Ordnance Survey maps is still valid.

Richard R. Oliver, 'The Ordnance Survey: a quick guide for historians', *The Historian*, 30 (1991), 16-19 is a short article written in the Ordnance Survey's bicentennial year and includes up-to-date references to recent scholarship.

W.A. Seymour (ed.), *A History of the Ordnance Survey* (Folkestone, Dawson, 1980) is an official history of the Ordnance Survey as a mapping institution.